Architect's Guide to

NEC4

Frances Forward

RIBA  **Publishing**

© RIBA Publishing, 2018

Published by RIBA Publishing, part of the Royal Institute of British Architects,
66 Portland Place, London, W1B 1AD

ISBN 978 1 85946 856 2

British Library Cataloguing-in-Publication Data.
A catalogue record for this book is available from the British Library.

Commissioning Editor: Alexander White
Project Editor: Daniel Culver
Production: Phil Handley
Typesetting: Academic + Technical, Bristol, UK
Printed and bound by Page Bros, Norwich, UK

www.ribapublishing.com

Contents

About the author

Frances Forward BA (Hons) Dip Arch MSc (Const Law) RIBA FCIArb is a practising architect, adjudicator, expert witness and lecturer with extensive experience in design, management and consulting roles within the construction industry, both in the UK and in Germany. Frances' Master's research in the early 1990s into a potential European standard form construction contract led her to explore the genesis of the fledging NEC contract and she went on to pioneer the use of the NEC on complex lottery-funded arts projects.

Frances set up her own architectural practice in 1999 and the NEC became her contract of choice, for both professional services and construction. In 2000, Frances was invited to assist the NEC Panel in drafting the NEC Partnering Option X12, and she subsequently became the architect member of the Panel during the drafting of the third edition, NEC3.

Frances expanded her architectural practice in 2005 to offer an additional role of management contractor to allow clients to procure both the design and the off-site fabrication/on-site construction of sustainable buildings from a single company. Her company model has matured into a specialist Passivhaus standard partnering supply chain, including both English and German specialist subconsultants and subcontractors, using NEC as the common form of contract.

Frances has contributed to several publications and conferences regarding the NEC and she lectures on the NEC at various University Schools of Architecture as part of their Professional Practice programmes.

Introduction

Over nearly three decades, the NEC has evolved from a 'revolutionary' new form of contract to become a mainstream standard form contract which is particularly conducive to collaborative working and contractual partnering. Use of the NEC has grown steadily in all sectors of the construction industry over its lifetime and a working knowledge of it has become essential for all architects, allied professionals, their clients and contractors involved in building projects. Following the OGC[1] endorsement of NEC3 on its publication in 2005, the use of any other form of construction contract on publicly procured projects declined to a point where no one involved in such work could afford to be ignorant of NEC3. The GCB[2] recommendation in 2017 that public sector organisations use NEC contracts and specifically NEC4 further underlines its influence. Use of the NEC is increasingly driven by client bodies and while architects in some specialist building sectors, for example, healthcare and transport, are relatively conversant with the NEC, other architects have limited awareness of it. Architects are also increasingly exposed to the NEC in the private sector and need to be comfortable with it in relation to all potential projects.

The RIBA Plan of Work and architectural education at Part 3 level acknowledge that architects give procurement advice; in practice, a lack of consistent knowledge about the NEC among the architectural profession could put architects at risk of giving incomplete procurement advice. The key aim of this book is to make an adequate level of knowledge readily available to all architects. The book is intended to make NEC4 accessible to many and at many levels, including those who have already built up a knowledge base with NEC3.

Given the legacy of the Victorian era on the evolution of older style standard form construction contracts, there is particular need for all professionals to look very clearly at the provisions of NEC4 as a 'modern' standard form contract, capable of supporting best practice in project management. Furthermore, given the historical divergence of the building and engineering sectors of the construction industry in the UK, there is also a need to distinguish the specific relevance of the multidisciplinary NEC4 contract to building sector professionals.

Encompassing the multidisciplinary nature of the NEC, a number of books have been written on various aspects and uses of the Contract; however, these have tended to focus on its use primarily in the engineering sector of the construction industry and on its legal interpretation. This book focuses in detail on the particular needs of the building sector within the construction industry and consequently sets the NEC4 scene in the context of an architect's perspective and professional responsibilities.

---

1   Office of Government Commerce.
2   Government Construction Board.

1 Background to the NEC

Procurement strategy

Contract typology

Conventionally, the parameters of time, cost and quality have been assessed in relation to choosing the correct type of contract for individual projects on the following basis:

- **time:** design and construction duration *and* certainty of end date

- **cost:** overall price (fees and construction) *and* certainty of final account

- **quality:** specification standards *and* workmanship on site.

The procurement analysis of the relative importance of time, cost and quality has historically led inexorably to a decision as to whether a traditional, a design and build, or a management procurement route is appropriate. However, such analysis has also long been predicated on the convention that time will be somewhat compromised under traditional procurement routes, quality will be somewhat compromised under design and build procurement routes and cost will be somewhat compromised under management procurement routes. The arguments leading to this convention are well rehearsed and need not be examined in detail here, as their only real relevance in the context of the NEC is that they represent an outmoded and arguably superseded approach to procurement strategy.

A further subset of contract typology is the payment mechanism, which conventionally falls into the following categories:

- lump sum

- remeasurement

- cost reimbursable.

These generic payment mechanisms remain relevant in the context of the NEC, albeit the NEC offers greater sophistication in their implementation than earlier standard form contracts.

It should be noted that no type of standard form contract, including NEC4, offers either a 'fixed price' or a 'guaranteed maximum price' (GMP) payment mechanism; these being inventions of those who seek to amend standard form contracts, or draft bespoke contracts, to highly polarise risk allocation.

Contract form

Professional drafting bodies historically published standard form contracts based on traditional procurement strategy[3] and subsequently responded to analysis of the so-called 'time/cost/quality' triangle by publishing additional design and build and management versions of their standard forms. Architects have long been used to providing clients with procurement advice and indeed are expected to advise on both a 'Project roles table' and a 'Contractual tree' at a relatively early stage in a project.[4] This important advisory role should take account of the need for flexibility and further review.[5]

Project-specific strategies

Increasingly, a need has developed for contracts to respond to individual project requirements in a more finely calibrated manner; project sponsors simply can no longer accept that only 'two-and-a-half' out of the three parameters of time, cost and quality are adequately controlled. The resultant requirement for project-specific procurement strategies leads to what might be described as a hybrid procurement route. Such a route inevitably calls for much more flexible contracts than conventional procurement routes and this may partly explain the apparent growth in entirely bespoke construction contracts being drafted for important projects.

There is arguably a fourth procurement parameter that most 21st-century construction projects are required to take into account and that is risk. NEC4 sets out to offer a highly flexible format which responds to the 'prototype' nature of many construction projects and provides the ability to build up an appropriate contract. NEC4 enables a breakaway from conventional procurement analysis and there is no necessary compromise between time, cost, quality or risk management.

3   Separation of design and construction.
4   RIBA Plan of Work 2013: Work Stage 1.
5   Ibid.: Work Stages 2–4.

Genesis and philosophy of the NEC

Origins

The genesis of the NEC[6] was an initiative in the mid-1980s by a new Legal Affairs Committee within the ICE[7] in London. This initiative resulted primarily from a general dissatisfaction with 'Victorian' style standard form contracts within the construction industry, which had been conceived of prior to the commonplace requirement for complex multidisciplinary projects and which had become increasingly convoluted, in response to the perception of a 'high-risk' and 'adversarial' construction industry. The initial strategy for a 'modern' contract was developed by a small team led by Dr Martin Barnes,[8] and a consultative version of the NEC was published in 1991; this was generally received with such enthusiasm that it was followed by an official first edition in 1993. The NEC received important endorsement in the UK Government/industry Latham Report[9] of 1994 and the NEC second edition was published in 1995. The partnering ethos of the NEC contract was further endorsed in the UK Government/industry Egan Report[10] of 1998. A review of the NEC in use and users' comments was undertaken by its drafting panel, under the auspices of its publishers,[11] and the third edition, NEC3, was published in 2005. Following extensive use of NEC3 by an expanding range of users, and a targeted review of those users' findings in contemporary practice, NEC4 was published in 2017.

Application – what is in a name?

An important factor in the interest generated in the NEC was its applicability to a broad range of 'engineering' projects. This was 'officially' extended to include all construction projects, following the Latham Report, although the revised title 'ECC'[12], intended to emphasise the range of applicability, never really captured end users' imaginations and the original name 'NEC' largely prevailed. Ironically, the initial interest of architects in the NEC might have been greatly increased, and subsequent interest accelerated, had the title been revised to 'Engineering and Building Contract'. The answer to 'what is in a name?' in this instance seems to be 'quite a lot'!

There was also a clear intention that the NEC should be conceived as a contract that would be operable globally[13] and the drafting is intended to facilitate diversity on a number of levels.

---

6   New Engineering Contract.
7   Institution of Civil Engineers.
8   BSc(Eng) PhD FICE FCIOB FAPM FICES MBCS CCMI FREng CBE.
9   Latham, M. (1994) *Constructing the Team*, London: HMSO.
10 Egan, J. (1998) *Rethinking Construction: Report of the Construction Task Force*, London: HMSO.
11 Thomas Telford Publishing, London.
12 Engineering and Construction Contract.
13 See Chapter 5.

Guiding principles

The NEC approach encompassed the concept that both the legal and the management requirements of a diverse range of 'modern' projects could be met in a single document and that the avoidance of 'legalistic' language would assist in that aim.

The principles of risk theory and risk management were also important considerations, and an early decision was made that the contracting parties and their representatives should be required to act in a 'spirit of mutual trust and co-operation'.[14]

The stated objectives of NEC are *flexibility*, *clarity* and *simplicity*, as well as providing a *stimulus to good management*. The key objectives in drafting NEC4 include providing a *greater stimulus to good management* and supporting *new approaches to procurement which improve contract management – evolution, not revolution*.[15] In practice, the NEC approach offers a range of benefits which are key to its success:

- 'pick-and-mix' contract conditions, to suit both the Project and the Project Team

- plain English, giving both legal and project management rights and obligations equal status

- real-time project management, with contemporaneous decision-making

- cross-industry application, facilitating multidisciplinary working practices.[16]

The NEC contract family

The NEC family relationship for architects

It is pertinent to emphasise that the NEC has been designed for extremely flexible use patterns and different family members will therefore have different levels of significance, depending on the disciplines of users. Architects will tend to be interested in all the family members (see Figure 1); however, they are likely to have the closest relationship with the NEC4 *Black Book* Engineering and Construction Contract and the NEC4 Engineering and Construction Contract Short Contract in the context of building contracts, and with the NEC4 *Orange Book* Professional Service Contract in the context of consultants' appointments.

The NEC4 Subcontract/Short Subcontract will also be significant for architects in the context of specialist design and installation. Historically, architects have tended to take much less notice of subcontract conditions than main contract conditions, often believing the detail of them to be a Contractor's responsibility and largely outside the sphere of an architect's influence. Given the decline in staff directly employed by contractors and the greater reliance on specialist subcontractors to realise projects, architects may ignore subcontract

---

14 NEC4 Core Clause 10.2.
15 NEC4 User Guide, June 2017.
16 Envisaged in the Banwell Report, 1964.

conditions at their peril. There have been recent attempts by several drafting bodies to improve coordination between main contracts and subcontracts; nonetheless, NEC contracts have been at the forefront of a coordinated approach since their inception.

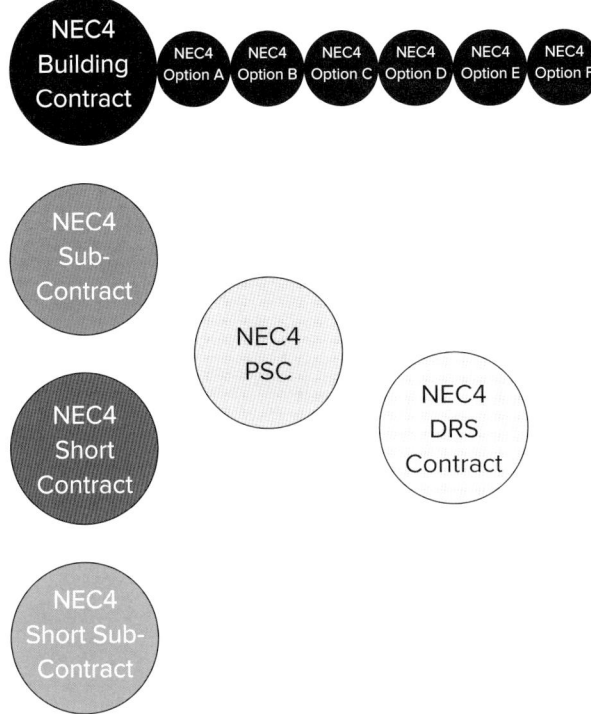

Figure 1: NEC4 'Immediate' Family: building projects

Compatibility, 'nesting' of contracts and uniformity

The risk of incompatible rights and obligations as between Client and Contractor within the building contract or as between the Client and Consultants within their professional services contracts is greatly reduced with the provision for back-to-back contractual arrangements. The risk of incompatible obligations as between Contractor and Subcontractor is similarly reduced. The back-to-back drafting has a further benefit, which is that there is no necessary connection between the status of the parties and the type of contract to be entered into, as between construction and professional services.

It is possible to 'nest' a number of contracts within each other, irrespective of whether the head contract is for construction or professional services. The conventional approach may be perceived as nesting subcontracts into a construction contract, or in the context of design and build procurement, also nesting professional services contracts into a construction contract. However, the flexibility of the NEC allows not only for the nesting of 'same' contracts within each other (e.g. NEC subcontracts or Professional Services Contracts), but also potentially for unconventional nesting such as construction within professional services.

Uniform application of the NEC methodology is encouraged by the publication in parallel with the contract versions of comprehensive User Guides to preparing and managing those contracts, as well as guidance on establishing appropriate procurement and contract strategies. These do not form part of the contract itself, but nevertheless provide valuable assistance in understanding and operating the project management principles of the contract.

NEC4 published contracts

Generally, there has been an attempt with NEC to provide as few separate contracts as possible, preferring the 'pick-and-mix' approach, although some separation has proved desirable in the evolution of the NEC family. Such separation remains unrelated to generic choice of procurement strategy as between traditional, design and build, or management, but rather relates to making the 'nesting' of contracts flexible and providing a clear interrelationship. All previously published NEC contracts were revised in 2017 as the fourth edition, and some additional contracts have been included in NEC4.

In order of their importance to architects, the NEC4 family comprises of the following publications:

The NEC4 'Building' (Engineering and Construction) Contract
(June 2017)

The so-called *'Black Book'*

Architects could be forgiven for unofficially renaming this the NEC4 'building' contract. It is very important for architects to bear in mind that this single book represents a standard form building contract blueprint, equally appropriate for traditional, design and build, management and hybrid procurements strategies. It is this universality that is of paramount importance in giving architects and their clients real choice, both at the outset of projects and, where necessary, during later stages of the procurement process.

The NEC4 'building' contract is also published individually for each of the six 'payment mechanism' main options A to F,[17] although this is for convenience, not necessity, and should certainly not be misunderstood, in that it remains a single contract form.

---

17   See Chapter 2, *Main option clauses.*

The NEC4 (Engineering and Construction) Short Contract

(June 2017)

(Dark Book)

Building clients may find this version of NEC4 appropriate where the Project is relatively straightforward, without the need for much fine-tuning of the contract conditions[18]. The Client and the Contractor communicate directly with one another, without a dedicated 'contract administrator', although there is provision for 'delegated authority' from the Client. Perhaps the most helpful way of deciding whether the NEC4 Short Contract might be appropriate is to consider it to be suitable for low-complexity, rather than low-value, projects. Historically, some standard form drafting bodies have made minor works contracts available and indicated that they are suitable for contracts up to certain, relatively low, monetary values. The critical point with the NEC4 Short Contract is that it may be eminently suitable for high-value contracts, provided that the work content of such contracts is relatively simple.

The NEC4 (Engineering and Construction) Subcontract

(June 2017)

The so-called *'Purple Book'*

It is no exaggeration to state that the only significant difference between the *Purple Book*[19] and the *Black Book* is simply that the *Purple Book* refers to the Contractor rather than the Client and to the Subcontractor rather than the Contractor. However, this deceptively simple swap is in turn the key to the success of the NEC4 Subcontract – it is genuinely back-to-back with the main contract *Black Book*.

The NEC4 (Engineering and Construction) Short Subcontract

(June 2017)

(Mauve Book)

In a mirror of the *Purple Book/Black Book* relationship, it is also no exaggeration to state that the only significant difference between the NEC4 Short Subcontract and the NEC4 Short Contract is simply that the Short Subcontract refers to the Contractor rather than the Client and to the Subcontractor rather than the Contractor. The Short Subcontract is therefore not only genuinely back-to-back with the Short Contract, but also offers an alternative to the *Purple Book* for subcontracted works of a simple nature under the *Black Book*.

---

18   See Chapter 2, *Secondary option clauses*.
19   See Chapter 4, Secondary Option X12 and *Subcontracting*.

The NEC4 Professional Service Contract (PSC)

(June 2017)

The so-called *'Orange Book'*

The PSC is interesting for architects, in that like the Subcontract, it offers the potential for back-to-back contractual arrangements.[20] This potential may be particularly relevant where architects are employed initially by Clients and subsequently by Contractors under design and build procurement arrangements, in that there is likely to be much less potential disparity between pre- and post-novation obligations. Another context in which the back-to-back potential is likely to assist architects is where they are involved in projects side-by-side with several other specialist consultants; whether the architects are acting as lead consultant or not, there will be much less risk of gaps and/or overlaps in the totality of the consultants' work.

The NEC4 Professional Service Short Contract (PSSC)

(June 2017)

(Bright Orange Book)

Architects may find this version of NEC4 appropriate when the professional services requirements are relatively straightforward and where building clients are lay people/consumers rather than professional/repeat clients.

The NEC4 Professional Service Subcontract (PSS)

(June 2017)

(Navy Book)

Architects may find this version of NEC4 appropriate when they are employing subconsultants to carry out professional services. It may also be appropriate for architects to be appointed when they are working for a Contractor to provide professional services under a design and build procurement route. This contract is a new addition to the NEC family with the publication of the fourth edition, NEC4.

The NEC4 Term Service Contract (TSC)

(June 2017)

(Grey Book)

Essentially, the TSC is intended to operate with the same flexibility and control as the *Black Book* for ongoing works, such as maintenance tasks, which cannot be fully defined from the outset.[21]

---

20 See Chapter 4, *Professional services.*
21 See Chapter 4, Secondary Option X12 and *Framework agreements.*

The NEC4 Term Service Short Contract (TSSC)

(June 2017)

(Pale Grey Book)

The TSSC offers a simple methodology for ongoing maintenance type works of a straightforward nature. The Client and the Contractor communicate directly with one another, without a dedicated 'contract administrator', although the Contract provides the option of the Client appointing an Employer's Agent.

The NEC4 Term Service Subcontract (TSS)

(June 2017)

(Ruby Book)

The TSS may be appropriate where a Term Service Contractor wishes to appoint Subcontractors for a period of time.

The NEC4 Design Build and Operate Contract (DBO)

(June 2017)

(Turquoise Book)

The DBO Contract is new to NEC4 and is intended to provide a vehicle specifically suited to clients who wish to appoint a Contractor with responsibility to design, construct and operate assets, whether the construction is new build or refurbishment.

The NEC4 Supply Contract

(June 2017)

(Red Book)

The NEC4 Supply Contract is intended for the purchase (locally or internationally) of high-value goods and related services, which may include design.

The NEC4 Supply Short Contract

(June 2017)

(Puce Book)

The NEC4 Short Supply Contract is intended for the purchase of relatively simple goods.

The NEC4 Dispute Resolution Service Contract (DRS)

(June 2017)

(Green Book)

The adjudicator has had a role under the NEC contract prior to the introduction of statutory adjudication in England and Wales.[22] It has always been considered advantageous to treat the adjudicator as a person involved in a project in a potentially positive way from the outset. The corollary to this perspective is that an NEC adjudicator should be signed up from the beginning, and that is the basis upon which the NEC4 Dispute Resolution Service Contract is intended to operate.

The NEC4 Dispute Resolution Service Contract is also the appropriate contract for appointing Dispute Avoidance Board members.

The NEC4 Framework Contract

(June 2017)

(Beige Book)

The NEC4 Framework Contract[23] provides a standard umbrella contract for other NEC contracts to be potentially instructed to pre-qualified suppliers[24] over a set period. Architects working in the context of public procurement[25] may find the NEC4 Framework Contract a valuable member of the NEC contract family, in that it obviates the need for bespoke framework contracts which can often conflict with the standard form terms under them.

---

22 Housing Grants Construction and Regeneration Act (HGCR Act) 1996 brought into force with the Scheme for Construction Contracts (England and Wales) Regulations on 1 May 1998.
23 See Chapter 4, *Framework Agreements*.
24 Whether Consultants or Contractors.
25 Within the European Union currently.

2 Structure and content of NEC4

'Pick-and-mix' assembly of the contract

Clause hierarchy and contract layout

Of paramount importance is the structure of the NEC4 contract conditions as a three-tier shopping list, comprising (1) core clauses, (2) main option clauses, and (3) secondary option clauses, from which the necessary items must be selected. The selection criteria are to be found in the project type and risk profile.

A fundamental drafting decision, which is key to the clarity of the NEC clauses in practice, was to take advantage of starting from the beginning and to arrange the document in a logical order. First, there is a clear group of concepts set out in nine 'core clause' sections and, therein, a clear sequence of clauses dealing with the specific nature of each concept. Cross-referencing between clauses is avoided and related elements are kept within the individual sections. The drafting of these core clause sections is genuinely generic, with the clear intention that they are applicable to any project, whatever its nature and wherever it is in the world, under whatever jurisdiction. Second, a single 'main option' is chosen to determine the pricing mechanism applicable to a particular project and to add the corresponding clauses required to operate it. Third, an assessment is made as to which, if any, of the 'secondary option' clauses will assist in fine-tuning the contract to meet the specific needs of that project. Finally, a decision is made as to which option applies for resolving and avoiding disputes.

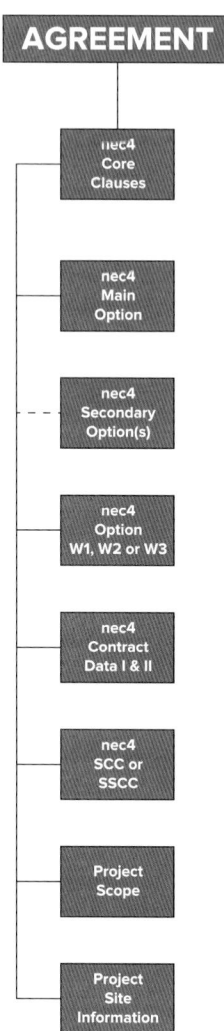

Figure 2: NEC4 Contract Structure

Necessary clauses

Many standard form contracts presuppose that all projects follow a similar enough route under a particular procurement strategy for all the contract conditions to be the same, i.e. generic. Even in the context of a clear decision as to the procurement route and therefore the contract form, this is inflexible and tends to favour a 'form-filling' mentality rather than a creative approach to assembling the appropriate contract for a certain project. There are parallels for this distinction in other areas of an architect's expertise, for example, when completing an NBS-style[26] specification, one approach would be to leave most standard clauses intact – just in case they prove useful, while the opposite approach would be to start from a blank sheet and only include those clauses which are considered strictly necessary. NEC4 is analogous to the latter approach, which tends to result in shorter documentation, avoidance of extraneous or superfluous clauses and consequential clarity.

'Designing' the project-specific contract

While some organisations have deliberately put in place standard contract preparation procedures to maintain quality, it is clearly inadvisable to follow such procedures when assembling the contract conditions of an NEC Contract. The whole point of the three-tier hierarchical structure is that it allows for the contract to be 'designed' as a tight fit for the needs of an individual project. It is not only acceptable, but also positively to be expected, that different main option and secondary option combinations will be chosen to augment the core clause sections for different projects. Even where projects are of a similar building type, or for the same client body, the decisions on appropriate options should be made afresh each time, to enable optimum performance of the NEC Contract on any one project.

NEC4 is a proactive contract, requiring hands-on management from the outset, including putting it together. There is no default version of the contract and it simply will not be operable if its assembly does not follow the envisaged structure or is incomplete. NEC4 will appeal greatly in the context of wanting a pragmatic framework within which to manage real projects effectively. The necessary skillset of architects is such that they should be well equipped to respond to the need to assemble the contract creatively, carefully and in adequate consultation with their clients.

---

26 National Building Specification.

Core clauses

There are nine core clause sections, each of which deals with an individual concept.

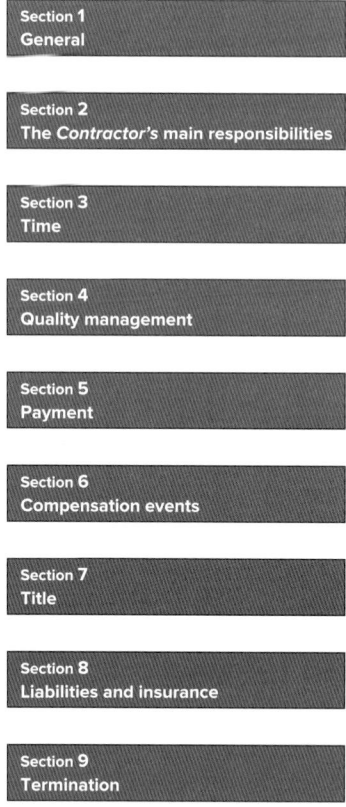

Figure 3: NEC4 Core Clauses

The following provides an interpretive commentary on the salient points within those concepts, rather than simply paraphrasing all the clause content.

Core Clause Section 1: General

This general section is key to operating the entire contract, and new users of NEC4 who have perhaps become used to skim-reading older forms of contract (whether because of familiarity or tedium!) would do well to be meticulous in working their way through this section. The drafting style throughout the contract is refreshingly succinct and the meaning and purpose of individual clauses should be relatively easy to discern. Ironically, there is more risk that the succinctness will be interpreted oversimplistically than there is that any clauses will be regarded as obtuse.

The first clause of the contract, Clause 10 Actions, has sometimes been regarded as its most controversial:

> 10.1 The Parties, the *Project Manager* and the *Supervisor* shall act as stated in this contract.
>
> 10.2 The Parties, the *Project Manager* and the *Supervisor* act in a spirit of mutual trust and co-operation.

In the early years of NEC, many probably regarded this clause as aspirational, rather than legally binding; now, it is much more likely that a failure to act as required might be construed as a 'breach of contract'. The statement in Clause 10.2 has come to be seen as virtually synonymous with the ethos of partnering. The progress of partnering has been such that it is no longer fanciful to ascribe legal obligation to that ethos.[27] The ethos is also consistent with the concept of 'good faith' under civil law jurisdictions, which is a well-established legal obligation in most European countries.[28]

In view of the NEC provision for use in different jurisdictions, it is important to check the law of the contract, whether in terms of what is appropriate when filling out the Contract Data, or in terms of retrospectively assessing legal obligations.

The explanation of identified[29] and defined[30] terms is extremely important to correct understanding and usage of all the following clauses, and this part of the contract will become well-thumbed by architects administering NEC4 contracts, whether acting as Project Manager, Supervisor or both.[31]

NEC4 has an 'entire agreement' clause,[32] which has the usual purpose of improving clarity by excluding any prior agreements. The Contractor is not expected to do anything illegal or impossible and must not do anything corrupt.[33] There is also a 'prevention' clause[34] which allocates the risk of unforeseeable events to the Client.

Types of communication are specified under NEC4,[35] both in terms of type and timescale – this is a critical factor in improving project management.[36] It is acknowledged that ambiguities or inconsistencies could exist in the contract documents and provision is made to resolve any such items.[37]

---

27 See Chapter 4, *Partnering*.
28 E.g. 'Treu und Glauben' under German Law.
29 In *italics* and identified in the Contract Data.
30 With capital initials and defined in Core Clause 11.2.
31 See Chapter 3, *People*.
32 Core Clause 12.4.
33 Core Clauses 17.2 & 18.
34 Core Clause 19.
35 Core Clause 13.
36 See Chapter 3, *Communications*.
37 Core Clause 17.1.

NEC4 envisages that the Contractor may need access to places that are not part of a building site as such and that there may also be reason to restrict access on a building site to specific areas – the combination of these being defined as the Working Areas.[38] This is a more sophisticated approach than simply giving a Contractor 'possession' of a site and in practice, allows better control and a manageable solution to any need to share spaces on a building site. There is provision for making additional areas available where this becomes necessary.

NEC4 focuses throughout on efficient project management and contemporaneous resolution of any matters that could affect time, cost or quality.[39] The early warning provisions and the maintenance of an Early Warning Register are vital provisions in this respect.[40]

Core Clause Section 2: The Contractor's main responsibilities

In Providing the Works,[41] the Contractor is only bound by the content of the Scope, which means in practice that the quality of the Scope is of paramount importance in controlling the realisation of any building being constructed under an NEC4 contract.[42] Architects would typically be responsible for the production and coordination of the Scope on building projects and will therefore be particularly interested in the relationship between its quality and that of the corresponding completed building.

The Contractor has potential responsibility for design of parts of a building, subject to individual project requirements.[43] In practice, those individual project requirements might suggest anywhere between 0% and 100% design by the Contractor, whether the percentage be of the entire building or of discrete constituent parts. The NEC4 *Black Book* building contract enables architects to have tremendous freedom in deciding exactly the right split in terms of what design work is allocated to whom and at what stage of the Project.[44] The Contractor also has potential responsibility for design of Equipment;[45] i.e. typically on building projects, temporary works necessary to realise the permanent works. The Scope is the correct place to accurately define any requirements for the Contractor to design any item, whether permanent or temporary. The Scope is also the correct place to identify the parameters of the Client's rights in relation to using any designs executed by the Contractor.[46]

---

38 Core Clause 11.2 (20).
39 See Chapter 3, *Dispute management*.
40 Core Clause 15.
41 Core Clause 11.2 (15).
42 See Chapter 2, *Scope*.
43 Core Clause 21.
44 See Chapter 3, *Design*.
45 Core Clause 11.2 (9).
46 Core Clause 22.

The Contractor has responsibilities concerning provision of key staff and any necessary replacement of staff. The Contractor also has potential responsibility to cooperate with Others,[47] including sharing the Working Areas. In practice, these responsibilities are extremely useful project management aids, particularly as they recognise the wealth of external bodies and stakeholders who can be involved in today's building projects.

The Contractor has potentially onerous responsibilities concerning any subcontracting,[48] although the correct discharge of those responsibilities reaps significant rewards in terms of efficiency and quality of outcome, which is arguably in both the Contractor's and the Client's best interests.

In view of the potential international use of NEC4, the Contractor's responsibilities in relation to health and safety are not stated in the core clauses but are to be stated in the Scope specific to an individual project. Architects building in the UK need to be aware of this distinction and ensure that appropriate references to relevant legislation[49] are made in the project-specific Scope.

Core Clause Section 3: Time

The temporal provisions of NEC4 are prescriptive, challenging and very good news. It is no exaggeration to state that architects have conventionally been in the hot seat in relation to delays on projects and that discharging their responsibilities with respect to extensions of time under many standard form contracts has become exceedingly difficult. It is an area where responsible architects would be justified in suffering sleepless nights, given the potentially disastrous combination of 'responsibility without authority' which appears to exist in the drafting of some standard form contracts in relation to assessing responsibility for delay. The nub of the issue is that if an architect is expected to make such assessments objectively and in accordance with the critical temporal path of a project,[50] then doing so without the benefit of an appropriate programme from the Contractor is well nigh impossible; producing such a programme independently, while possibly sensible if shown to the contracting parties, is nevertheless illogical.

Where NEC4 differs radically from many other standard form contracts is in its approach to the provision by the Contractor of a programme which can be seen to assist everyone involved in a project in achieving clarity and certainty of temporal outcome, both from the outset and as the Project progresses.[51] The status of that programme, once initially accepted and subsequently reaccepted by the Project Manager,[52] is binding on both parties to an NEC4 building contract, i.e. Client and Contractor alike.

---

47   Core Clause 11.2 (12).
48   See Chapter 4, *Subcontracting*.
49   The Construction (Design and Management) Regulations 2015.
50   *Henry Boot Construction (UK) Limited v. Malmaison Hotel* 2000.
51   Core Clauses 31 & 32.
52   Core Clause 11.2 (1).

NEC4 distinguishes between the starting date and the first access date, thus catering for the eventuality of early Contractor involvement, prior to commencement of construction on site.

The conventional distinction between a contractual completion date and the actual date of completion is made in NEC4 as well, the former being stated in the Contract Data and the latter being certified by the Project Manager.

NEC4 provides for Key Dates, which in practice can be very helpful on some relatively complex building projects where a Client may need to ensure critical dates along the way are controllable, without necessarily requiring Sectional Completion.[53]

A novel feature for architects to find in a standard form contract is the ability in NEC4 to propose acceleration.[54] This provision should not be used lightly and clearly has potential financial implications; however, there will be building clients and building projects where there would be undoubted benefit in the ability to suggest 'official' acceleration to achieve earlier completion of a project, or part of a project, and to formalise this if the Contractor agrees.

Core Clause Section 3 is one of the most fundamentally different concepts relative to other standard form building contracts; it is also one of the most rewarding sections of NEC4 for architects to get to grips with.[55]

Core Clause Section 4: Quality management

The role of the Supervisor will be new to architects using NEC4 for the first time.[56] The responsibilities in relation to quality are not merely advisory, as is often the case with a 'clerk of works' role; they are an integral part of the overall contract administration to be carried out on behalf of the Client.

The multidisciplinary nature of NEC4 is such that foreseen testing and inspection of specific items is envisaged as a potential project requirement and this is provided for on the basis that it applies to any parts of a project so designated in the Scope.[57] Historically, some building projects have not had any requirement for foreseen testing or inspection; nevertheless, there are now many contexts where the clear contractual provision will benefit projects (e.g. concrete cube testing for crushing strength, visual inspection of full scale mock-ups of facade materials, air-tightness testing for building regulations compliance, etc.). It is important at design stage that architects give appropriate thought to necessary inspection of samples and testing requirements, in order that appropriate specification can be incorporated into the Scope.

---

53 Secondary Option Clause X5.
54 Core Clause 36.
55 See Chapter 3, *Programme*.
56 See Chapter 3, *People*.
57 Core Clauses 40.

Core Clause Section 4 also covers dealing with unforeseen testing and inspection, i.e. checking for defects. There are reciprocal obligations as between the Supervisor and the Contractor to notify each other of each defect as soon as they find it.[58]

Core Clause Section 5: Payment

The operation in practice for the payment provisions within NEC4 will depend heavily on which main option is chosen. The primary reason for this is that both the definition of Defined Cost[59] and the definition of the Price for Work Done to Date (PWDD)[60] are different for each main option, as is to be expected in the context of each main option allowing a different payment mechanism. Architects initially observing projects under the NEC4 form of contract may find it helpful to look at the main option payment provisions simultaneously with the core clause payment provisions – this is essential for any architect acting as Project Manager and therefore administering those payment provisions! It is also important to realise that the payment provisions under each of the main options rely on the proper execution of the relevant parts of the Contract Data in order to become fully operable. It is exceedingly rare under NEC4 for payment difficulties to arise if the contract has been assembled properly; however, if there has been any lack of rigour in doing so, it is fairly predictable that severe difficulties and unexpected outcomes can arise.

Architects should take particular note of the requirement to account for applicable tax, in addition to substantive sums, in amounts due for payment, as this is not usual in other standard form building contracts. Architects should also note, where applicable, the requirement to account for interest in the case of any corrections to substantive sums.

There is a deliberately draconian provision in the payment section which effectively withholds 25% of the amount to be paid to the Contractor until such time as a first programme is submitted by the Contractor.[61] This is intended to severely disincentivise the Contractor from failing to fulfil the programme requirements in Core Clause Section 3 (Time),[62] because such failure would in turn make other provisions within the contract, notably the compensation event provisions, inoperable.[63]

Core Clause Section 6: Compensation events

This section is regarded by many as the magnum opus of the NEC core clauses. Section 6 is certainly a section that received many comments in relation to earlier editions of NEC and it was probably the most amended section between the second and third editions of NEC.

---

58 See Chapter 3, *Defects*.
59 Main Option Clauses A11.2(23), B11.2(23), C11.2(24), D11.2(24), E11.2(24) & F11.2(25).
60 Main Option Clauses A11.2(29), B11.2(30), C11.2(31), D11.2(31), E11.2(31) & F11.2(31).
61 Core Clause 50.5.
62 Core Clause 31.1.
63 Core Clause 63.5.

Essentially, the compensation event procedures in Core Clause Section 6 of NEC4 are intended to deal simultaneously with both the financial and the temporal consequences of foreseeable events which materialise during the course of the contract and which are at the Client's risk.

The whole idea of 'compensation events' is pivotal to the successful operation of NEC. In the context of risk management, it is important to acknowledge that all risks which materialise on a project are allocatable to one or other party to the contract for that project. The well-rehearsed, albeit somewhat glib, risk management advice that residual contractual risk should be allocated to the party best able to carry it is completely compatible with the concept of compensation events. The contractual list of compensation events,[64] while specific and exhaustive, is a powerful tool in appropriate risk allocation and risk management on an individual project. Anyone questioning how a prescriptive, generic list of compensation events can be flexible enough to manage a plethora of varying risk profiles on different projects would be missing a key feature of NEC4: i.e. the author of the Scope is in virtually complete control of the Project's destiny! Notifying, assessing and implementing compensation events all take place relative to that project-specific Scope and the Accepted Programme and outcomes will therefore be as variable as the risk profiles and Scope and Accepted Programme content for individual projects.

By inextricably linking time with money and by relating both to objective contract documents, i.e. the Scope and the Accepted Programme, it is possible through correct operation of Core Clause Section 6 to maintain throughout any project contract both a running final account and a predicted date for completion, both of which will normally[65] be a maximum of three weeks behind real time.

Core Clause Section 7: Title

On most projects, Core Clause Section 7 probably attracts the least attention of all the core clause sections; certainly its content and brevity indicate that architects can distil the salient points relatively quickly.

The concept of Working Areas is helpful to architects in giving clarity over ownership of, and right to payment for, materials intended to be included in the works,[66] which are not yet on the actual building site. Conversely, there should be no question of the Contractor expecting payment for any materials stored outside the Working Areas, unless the NEC4 contract has been executed to include Advanced Payment.[67]

---

64 Core Clause 60.1.
65 Core Clause 62.3.
66 As identified in Contract Data Part One.
67 Secondary Option X14.

This section also ensures that the Client is not left with unwanted temporary items at the end of the Project[68] and that there is clarity over items found within the site during construction.[69]

Core Clause Section 8: Liabilities and insurance

The respective generic risk allocation as between the Client and the Contractor is set out in this section.[70] This should be distinguished from the project-specific risk allocation which is largely defined by the content of the Scope.

The default is that generic liabilities are required to be covered by insurance and that insurance is taken out by the Contractor in joint names, unless the Contract Data specifically gives any insurance obligations to the Client. There is an insurance table[71] summarising the type of cover required, which falls into the four categories of: (1) works insurance, (2) equipment (temporary works/machinery) insurance, (3) public liability (persons and property) insurance, and (4) employer liability insurance. The Contract Data is the correct place to identify any swap of insurance obligations from Contractor to Client, as well as any additional insurance requirements, for example, professional indemnity insurance.

Given the potential for NEC4 to be used in different countries and jurisdictions, there is a need to ensure that both types and levels of insurance cover are in accordance with the applicable national law and that the Contract Data is filled out accordingly.

There is a reciprocal requirement for whichever party is responsible for taking out the insurances to submit insurance certificates and/or policies to the other party for acceptance.[72] There is also a reciprocal right for a party to counter-charge the other party for having to take out insurance in the absence of proof from that other party, if they are responsible for taking out such insurance, that the required policy is in place.[73]

Core Clause Section 9: Termination

This is perhaps the least easy core clause section to digest, in that it sets out specific rules to be followed in the event of termination by either the Contractor or the Client. These rules are relatively complex, and the only consolation is that this entire core clause section will be relatively rarely used.

---

68 Core Clause 72.
69 Core Clause 73.
70 Core Clauses 80 & 81.
71 Core Clause 83.
72 See Chapter 3, *Communications*.
73 Core Clause 86.

Either party may terminate for the reasons[74] set out in the Termination Table,[75] which correlates the procedures to be followed[76] and the payment amounts due on such termination.[77]

Main option clauses

There are six main options, a single one of which must be chosen in order to provide a payment mechanism and make an NEC4 contract operable.

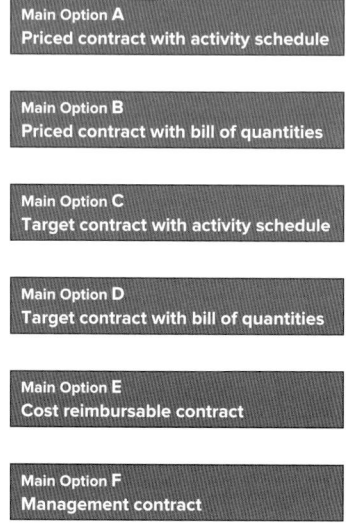

Figure 4: NEC4 Main Option Clauses

It has sometimes been questioned whether NEC contracts are fairer, i.e. have a better balanced allocation of risk, than other standard form contracts. The only correct answer to such a question is that 'it depends'. One of the key things it depends upon is a rational choice of the contractual payment mechanism, that is, the one, and one only, main option. Risk allocation can be significantly altered on the same project by virtue of the choice of main option. New users of NEC4 should not treat the main options as a pick and mix where they necessarily have to try all the flavours! It is entirely to be expected that professionals and their clients working predominantly in one sector of the construction industry will not go through the whole array of NEC4 main options. Architects working not only on building projects, but also possibly specialising in a certain building type with a specific client base, may choose to use the same main option on numerous projects. This is most certainly not

---

74 Core Clause 91.
75 Core Clause 90.2.
76 Core Clause 92.
77 Core Clause 93.

because NEC4 should be 'standardised' in a practice, but because it is foreseeable that when proper procurement analysis takes place on each project, and each project has similarities, then it follows that the same NEC4 main option may consistently provide the best contractual fit for a particular project.

In summary, the choice of main option is crucial to the successful operation of an NEC4 contract. Architects should keep an open mind in analysing the right choice for any one project but should not be tempted to choose a different main option 'just for a change'.

Main Option A: Priced contract with activity schedule

Main Option A essentially creates a lump sum contract, suitable for a range of building projects. Most architects experienced in administering other standard form contracts on a traditional procurement route will probably find Option A the easiest to relate to; however, NEC4 Option A is not restricted to traditional procurement. After carrying out a full analysis of procurement strategy for a particular building project, architects should not be concerned if they regularly decide that Option A is the most appropriate NEC4 main option to put forward to their client.

The payment mechanism under Main Option A is incredibly straightforward, in that the Contractor is entitled to be paid for each completed activity. It is important to realise the significance of activities having to be complete. First, complete does not mean nearly finished, it means 100% finished; the Contractor is thus incentivised to stay on programme. Second, there is no need for either debate or specific 'valuation' expertise in order to determine the value of a completed activity; it will simply be the monetary value ascribed to the relevant activity within the Contractor's Activity Schedule. This is why quantity surveyors need not perform a conventional valuation role in relation to NEC4 Option A contracts.

Those who remain unfamiliar with activity schedules under any standard form building contract sometimes question why an NEC4 activity has to be 100% complete. The answer is simply that there is no mechanism for assessing the value of partially completed activities. While it might be very easy and relatively accurate to pro-rate, for example, five courses of a 10m length of blockwork which is to be 15 courses high and the same length when complete, it would clearly be much more difficult to pro-rate, for example, a partially completed mechanical or electrical services installation. The benefits of allowing a Contractor to split up a project into activities relating to appropriate work elements, in relation to both logical progress on site and cashflow, would be completely lost if the Client still had to ensure that some sort of pricing document be produced in tandem, solely to enable partially completed work elements to be valued. Indeed, there would be significant scope for disputes to arise due to potentially conflicting methods for assessment of value.

If difficulties do arise with NEC4 Option A, it is likely to be due to a lack of understanding of the principles of activity schedules per se, rather than the requirements of NEC4. It is perhaps significant that while a number of standard form contracts have offered activity schedules as an alternative payment mechanism for some time, there is relatively little anecdotal evidence of their use in the building sector of the construction industry, and perhaps even less evidence of their success where occasionally used under other standard form building contracts.

It may be helpful to consider some particular criteria which contractors should give due regard to when splitting up a project to prepare an activity schedule:

- Does the Activity Schedule cover the entire works? That is, does there need to be a catch-all activity for 'everything else' other than the listed activities, or can the list of activities be exhaustive? 'Missing' activities could clearly lead to difficulties over entitlement to payment.

- Upon what basis does a project lend itself to be split up? For example, trade-related elements, demarcated geographical areas on the site, etc. In practice, this will depend on the size and nature of the particular project; many projects may benefit from being split up on a two- or three-tier basis. There is no magic right number of activities on any project; too few activities could lead to cashflow difficulties and too many could become an administrative burden. It may be advantageous to management clarity to group activities which are interdependent.

- What is the relationship between physical activities and management activities? That is, are there benefits in separating them or are they inextricably linked?

- Items which are conventionally priced under the heading of 'Preliminaries' are still activities and should appear in an activity schedule accordingly. This is essential under the NEC4 form of contract because of the NEC principle of inextricably linking time with money.

- Are the individual physical activities logically related to technical requirements in executing the works on site? For instance, if a concrete pour has to be done in two stages because of temporary works obstructions, then it should clearly also comprise of two separate activities.

- Have the durations of activities been considered in relation to contractual payment intervals? For example, if all the substructure concrete is to be poured sequentially, but that sequence will last for five weeks and the payment interval is four weeks, then it will assist cashflow if that physical sequence is split into two activities.

It is sometimes questioned how the NEC4 Option A provisions should work in the case of a Contractor realising either that an activity cannot be completed as a discrete element of work, or that they wish to change their methodology for an activity. It is also often questioned how the NEC4 Option A provisions should work in the case of a change to the required work under the Contract affecting one or more activities. The key to answering these points is to understand the relationship between the Activity Schedule and the Accepted Programme.

In the case of the Contractor wishing to change the sequence of work on site, they are entitled to do so, but they must alter the Activity Schedule and resubmit it, if necessary with a correspondingly revised programme.[78]

In the case of an instruction to change the works required under the contract (the Scope), the Contractor is obliged to change the Activity Schedule accordingly.[79]

For routine payments as a project progresses, Main Option A simply requires the Project Manager, possibly with assistance, to be capable of assessing whether any activity is 100% complete and should therefore be included in the certified sum for payment. In practice, any architect who has been involved in the preparation and coordination of the design documentation and who has studied the Contractor's Activity Schedule should have no difficulty in deciding when an activity is 100% complete.

In addition to routine payments, the Main Option A payment mechanism also has to deal with payments that may become due as a result of instructed changes under the contract. Main Option A provides for a fee to be applied in the case of compensation events, that fee being a percentage which is applied to the Defined Cost of the work.[80]

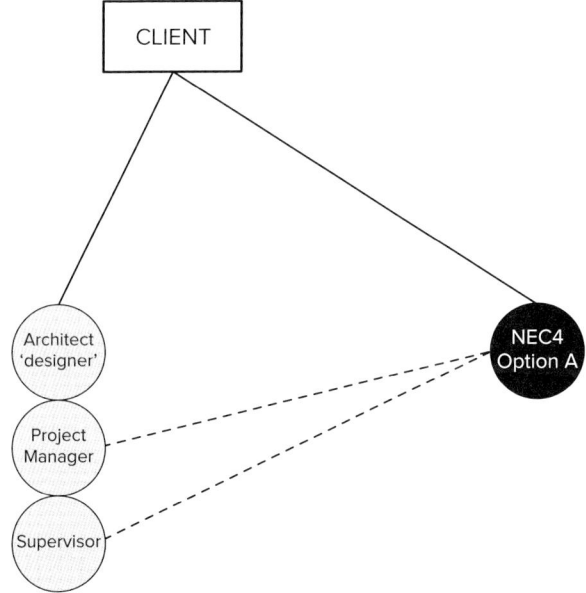

Figure 5: NEC4 Main Option A: Architect Acts as PM and Supervisor

78   Main Option A Clause 55.
79   Main Option A Clause 63.14/16.
80   Core Clause 11.2(10).

Main Option B: Priced contract with bill of quantities

Main Option B effectively creates a remeasurement contract because of the minimal inaccuracy margin permissible in the Bill of Quantities, without it becoming grounds for a compensation event. If architects are putting Main Option B forward to their clients as the most appropriate payment mechanism for a project, they should emphasise that it does not create a lump sum contract and that the potential risk for quantity changes would rest with the Client.[81]

This main option is possibly the one which many architects who are experienced in administering other standard form contracts on a traditional procurement route will tend to gravitate towards, if only because of the familiarity of the terminology 'Priced Contract with Bill of Quantities'. In practice, this could cause difficulties for architects and their clients alike, unless there is complete clarity over the issue of remeasurement, i.e. complete understanding and acceptance of the fact that because Main Option B does not automatically create a lump sum contract, even if no changes were instructed, the price would not necessarily remain fixed.

This is not a defect in NEC4, but rather a natural consequence of the NEC principle of flexibility and its design as a contract which caters for a potentially vast range of project requirements. Any architects who have worked on multidisciplinary projects such as rail infrastructure are likely to have come across the custom of remeasuring engineering quantities. Equally, architects who have worked on refurbishment projects are likely to be familiar with standard form building contracts that are either 'without quantities', or 'with approximate quantities', i.e. contracts that assume measurement, or remeasurement, of quantities.

Main Option B envisages conventional preparation of a Bill of Quantities on behalf of the Client at tender stage and it foresees the possibility that individual lump sums may be allocated to specific items which are included in the scope of work to be carried out by the Contractor. Option B provides a payment mechanism where the Contractor is entitled to be paid for each item in the Bill of Quantities on the basis of quantity multiplied by the rate and is entitled to be paid for lump sums. This accounts for why clients will normally retain quantity surveyors or cost consultants in relation to NEC4 Option B contracts, both for preparation of the Bill of Quantities pre-contract and to perform a 'valuation' role post-contract.

One of the critical differences between NEC4 Main Options A and B in the context of building projects is that the fees for preparation of the Option B Bill of Quantities will be borne directly by clients and the pre-contract programme will have to allow adequate time for its preparation, whereas the Option A Activity Schedule will be funded by tendering contractors as part of their bidding risk. While Option A tender periods should reflect the Activity Schedule production time, this is unlikely to increase the length of a pre-contract programme by as much as the Option B Bill production.

---

81   Main Option B Clause 60.4.

One slight philosophical anomaly between Main Option B and Option A is in the context of any individually allocated lump sums under Option B relative to activities under Option A. The Option B payment mechanism allows for proportional payment for lump sums on the basis of proportional completion of each lump sum. This is in direct contrast to the Main Option A payment mechanism of avoiding proportional payment and only allowing payment of 100% complete activities. Main Option B does not, therefore, assist in incentivising the Contractor to stay on programme in the way that Option A does. It is also foreseeable that a dispute could potentially arise under Main Option B about the precise proportion of a lump sum which had been completed at the end of a particular payment interval. While it is consistent with the whole principle of a Bill of Quantities that proportional payment takes place, it is tempting to conclude that Main Option B is much less radical than Option A in its approach to incentivisation and dispute avoidance techniques. To use the 'carrot and stick' analogy, Option A seems to proffer plenty of carrots, whereas Option B seems to still rely more on sticks.

If using Main Option B, perhaps the moral is to aim to limit lump sums to items that can easily be proportioned, as well as to ensure that the accuracy of quantities is meticulous. However, in the context of all but the simplest of building projects, the ubiquitous 'M&E' installation is likely to present a challenge in relation to both of those aims.

In addition to routine payments under Main Option B, the payment mechanism also has to deal with payments which may become due as a result of instructed changes under the Contract.[82] As with Main Option A, Option B provides for a fee to be applied in the case of compensation events, that fee being a percentage which is applied to the Defined Cost of the work.

---

82   Main Option B Clause 63.15.

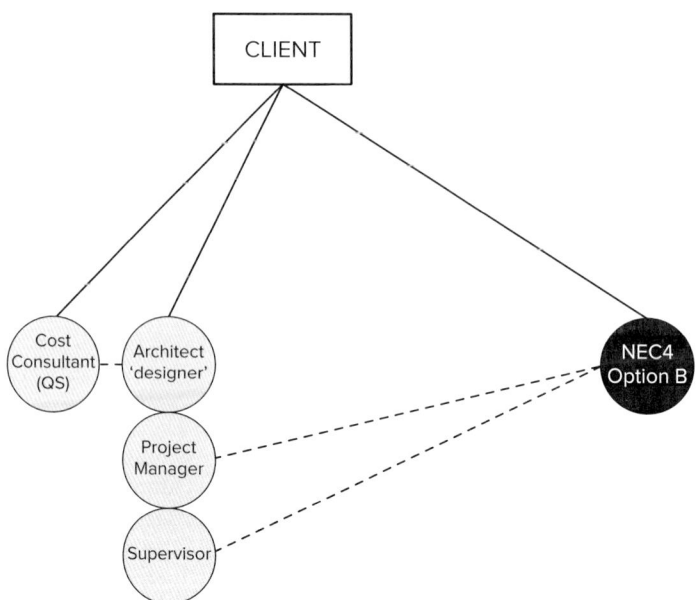

Figure 6: NEC4 Main Option B: Architect Acts as PM and Supervisor

Main Option C: Target contract with activity schedule

A target cost is agreed at the outset, with an incentive to the Contractor to achieve or to better the target, by offering a share in any savings. The cost of the Project is distributed into an Activity Schedule prepared by the Contractor, in a manner similar to Main Option A. The Activity Schedule under Main Option C is perhaps best understood as a distribution of the costs making up the Target Cost, where the sum of the constituent parts (activities) may or may not be greater than the whole Target Cost, depending on what the Contractor actually spends in terms of Defined Cost. Conversely, the sum of the constituent parts (activities) under a Main Option A Activity Schedule equals the whole contract lump sum. This distinction has an important impact on risk profile and incentivisation. It is one of many examples in NEC4 of where the status and implementation of things (whether documents such as an Activity Schedule or Defined Terms such as PWDD) is deliberately somewhat different depending on which main option has been chosen.

Target incentivisation

Note that with Main Option C, the allocation of risk can be significantly weighted by the share percentages allocated to each party to the NEC4 contract – hence the expression 'pain/gain' share. The intention is that the proportions are agreed, rather than imposed, as much of the potential benefit will be lost if excessive risk is priced as a contingency item which may or may not materialise. It would not be unusual for the share percentages for any pain and any gain to be unequal, subject to a proper assessment of which party is best able to carry certain project risks.

While target cost procurement strategies are not unique to NEC4, an architect's first encounter with target costs may well have been, or will be, via NEC. There are now a number of examples of 'repeat' clients using Main Option C contracts, both in order to better performance relative to other procurement strategies and in order to compare performance between their own projects and supply chain teams.[83]

In implementing a target cost strategy, the first long-standing concept to relegate is that of having a contract sum – there isn't one! There is only the Target Cost calculated for that individual project. An obvious, but sadly sometimes overlooked, factor in the success of target cost contracts is the accuracy of the cost plan upon which the Target Cost is based. Typically, most clients contemplating a target cost procurement strategy will be experienced clients, building projects of relatively large size, above average complexity and sensitive accountability for spending. Such clients often have experienced in-house procurement advisors or seek sophisticated procurement advice from consultants, which has both advantages and disadvantages. If the calibre of the advice given is high, there will undoubtedly be an emphasis on detailed risk analysis and risk management, which in turn are likely to lead to a careful setting of both the Target Cost and the pain/gain share percentages.

However, experience over the lifetime of NEC contracts suggests that very occasionally a somewhat 'unreconstructed' approach to risk management is adopted, whereby a client may be encouraged to shunt as much risk onto a contractor as possible. This way of thinking may have had its origins in older, more adversarial standard form contracts and may still be considered appropriate in some circumstances. Nonetheless, if it were to lead to a client managing to sign up a contractor on an NEC4 Option C contract with an artificially low Target Cost and a share percentage of 100% in any overspend relative to that target, it does not take much imagination to contemplate that the outcome might be very painful indeed, not just for the Contractor, but also for the Client.

---

83 E.g. healthcare and food retail sector clients.

The key point is that starting on the basis of the right Target Cost should be an even higher priority than agreeing the right share percentages for the risk of deviation from that cost. One of the most obvious ways of improving the accuracy of the target is to involve the Contractor(s) bidding for the work, in that they ought to have access to at least as accurate figures as cost consultants. If the list of contractors bidding for a job has been drawn up appropriately, such contractors ought to be in a strong position not only to contribute to accurate cost forecasting, but also to critique design solutions and offer useful value engineering input. This is not an NEC-specific point; however, an NFC4 Option C contract will have a massively improved chance of success if there really is 'a spirit of mutual trust and co-operation' between Client and Contractor teams with regard to the design of the contract as well as the building.

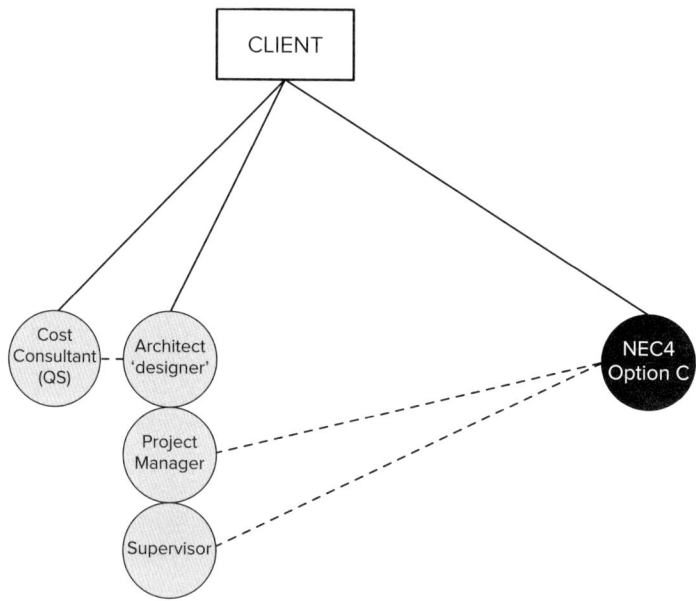

Figure 7: NEC4 Main Option C: Architect acts as PM and Supervisor

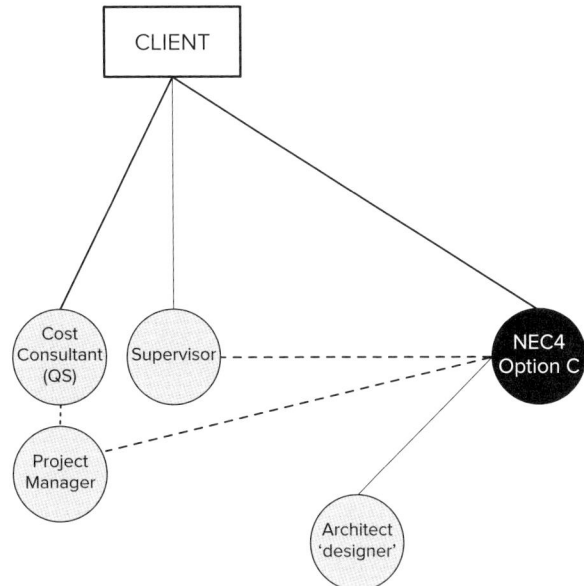

Figure 8: NEC4 Main Option C: Architect employed by D&B Contractor

Main Option D: Target contract with bill of quantities

As with Main Option C, the incentive to achieve the target relies on the requirement that any overspend is shared in a pre-agreed proportion between the parties.

With Main Option D, as with Option C, the fairness of the so-called 'pain/gain' share relies on both the accuracy of the cost planning to arrive at a realistic Target Cost and on the share percentages and share ranges agreed to form part of the contract.

With Main Option D, as with Main Option B, the minimal inaccuracy margin permissible in the Bill of Quantities, without it becoming grounds for a 'compensation event', effectively creates a remeasurement contract.

Looking at the payment mechanism envisaged with Main Option D in relation to building projects, there is clearly a need to assess risks associated with the accuracy of the Bill of Quantities, the accuracy of the cost plan upon which the Target Cost is based and finally the pain/gain share percentages. The overall risk profile that may emerge on a project under Main Option D could perhaps be a step too far for many architects and their building clients, without spending a disproportionate amount of time on pre-contract analysis and therefore a disproportionate amount of money on cost consultancy fees. This is not so much a criticism of the payment mechanism represented by Main Option D, but instead a recognition of the vast number of project types and sizes for which NEC4 is intended to cater. There follows a consequential acknowledgement that not all of the main option payment mechanisms will be equally suited to all project types or sizes.

The following is an example of a building project risk profile which might benefit from using Main Option D in terms of high motivation of both client and contractor to achieve a well-managed outcome:

- a large commercial project which includes infrastructure elements

- an experienced client wishing to keep control of the pricing document, i.e. the Bill of Quantities

- potential for early contractor involvement which could lead to a negotiated Target Cost

- acceptance that quantities of many items would be remeasured

- incremental pain/gain share percentages, i.e. where any spending over or under the Target Cost is not penalised or rewarded absolutely, but in proportion to the extent of deviation from the Target Cost, upwards or downwards

- desire for reciprocal pain/gain share percentages, i.e. not necessarily entirely equal share percentages for the Client and the Contractor as between 'painful overspend' and 'gainful underspend', but a correlation between the share percentage allocated to a party and that party's ability to manage the risk.

There is anecdotal evidence among NEC4 users that Main Option D has been significantly less used on building projects than Main Option C, where an incentivised Target Cost is desirable. This is perhaps encouraging as it indicates that NEC4 users have become quite discerning about risk management on building projects.

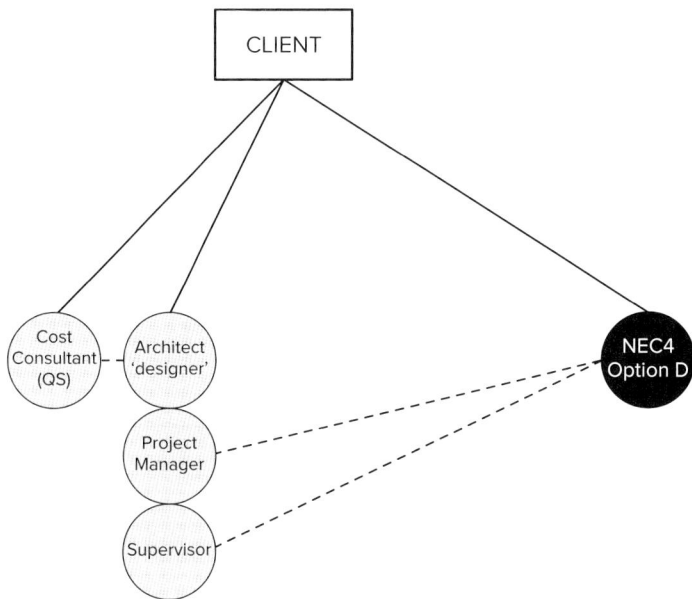

Figure 9: NEC4 Main Option D: Architect Acts as PM and Supervisor

Main Option E: Cost reimbursable contract

An 'open book' accounting policy is intended to operate, with the Contractor being paid at Defined Cost,[84] plus a percentage fee for profit and overheads.

Open book accounting is a concept which has been completely embraced in some quarters of the construction industry and yet is still regarded with a degree of disbelief in other quarters. The idea of disclosing actual invoices and declaring profit and overhead percentages also seems to find more favour in some countries than others. Ultimately, comfort with a payment mechanism which is based upon open book accounting is as much a matter of culture as it is of contractual arrangements.

---

84   Option E Clause 11.2 (24).

Main Option E is predicated on the Contractor being able to recover costs and overheads and make a profit and there will be some building clients whose reaction to this approach might be 'Why should contractors carry no risk?' However, experience should tell the building industry that 'risk shunting' in the opposite direction rarely benefits clients, as contractors who are expected to gamble in terms of their ability to make any profit will try to protect their interests. This is not an NEC4 phenomenon, but merely human nature playing its part. With any building contract, unless there is a payment mechanism which acknowledges profit as a right for contractors, contractors will tend to attempt to build in contingencies of some sort as a way of protecting their legitimate expectations in carrying out work for clients.

As envisaged with any cost reimbursable type contract, NEC4 Option E permits an early start on site, with relatively little final information available to the Contractor. While this represents a risk to the Client in terms of outturn cost on a building project, the decision to use any of the NEC4 main options should be founded on sound procurement analysis.[85] It may be that the Client has set time as a key parameter on a particular project, in which case Main Option E may offer advantages.

The manner in which an element of building work is costed under Main Option E is independent of the timing of the instruction to carry out that element of work. In other words, because of the way in which payment at intervals is assessed under Main Option E, it makes no difference to the calculations whether the assessment includes items which were part of the contract at the outset, additional items which have been instructed post-contract, or a mixture of the two. Nonetheless, because of the inextricable link under NEC4 between time and money within the compensation event procedure, the contract programme will clearly be extended by significant post-contract instructions. Architects need to weigh up carefully in their procurement analysis with their clients how best to control not only the parameter of time relative to the other parameters, but also the potential benefits of an early start on site – which Main Option E is suited to – relative to a longer duration on site which could arise from compensation events.

---

85   See Chapter 1, *Procurement strategy*.

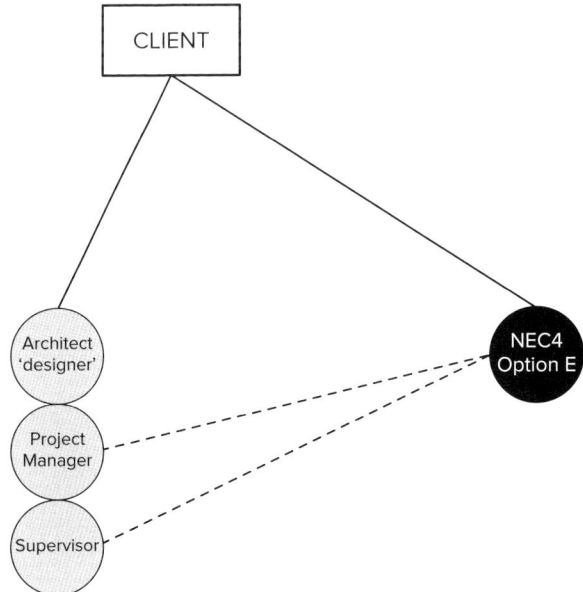

Figure 10: NEC4 Main Option E: Architect Acts as PM and Supervisor

Main Option F: Management contract

There is anecdotal evidence that this main option has been the least used of all the NEC main options across all sectors of the construction industry. Perhaps the reason for this is simply that the other five main options already offer significant flexibility of payment mechanisms and NEC4 as a whole offers significant tailoring of procurement routes; i.e. a management style procurement route can be achieved under NEC4 without necessarily incorporating Main Option F.

The intention with Main Option F is that the Contractor manages the subcontract packages and prepares forecasts of the total Defined Cost[86] of all work and receives this plus a percentage fee.

Possibly the best way for architects to assess the relative benefits of Main Option F when advising their clients on the assembly of an NEC4 contract is to focus on the degree of subcontracting desired and on how well prepared the scope of those subcontract packages is likely to be at the point when the NEC4 contract is to be entered into.

---

86   Option F Clause 11.2 (25).

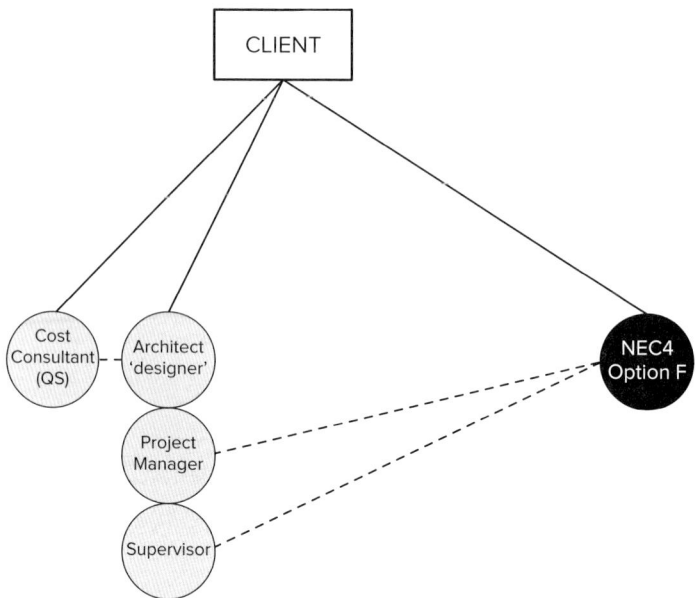

Figure 11: NEC4 Main Option F: Architect Acts as PM and Supervisor

Secondary option clauses

The secondary options can be operated on a pick-and-mix basis in any combination to tailor the contract as closely as possible to the needs of the project (see Figure 12).

These secondary options broadly fall into two categories:

- **Options which introduce choice in standard procedures.** Those options containing processes that conventionally have been included in the contract conditions of standard form contracts, whether needed or not, i.e. without giving a choice. An example of such a process would be Secondary Option X16: Retention.

- **Options which introduce additional standard procedures.** Those options containing processes that conventionally have not been available in standard form contract conditions without the introduction of bespoke drafting. An example of such a process would be Secondary Option X3: Multiple currencies.

It is important to note that there is no obligation to introduce any of the secondary options; an NEC4 contract is operable with none of them included. Conversely, most of the secondary options are not mutually exclusive to one another. That is, in theory, most of the secondary options could be included together in an NEC4 contract, the exception being that Secondary Option X20: Key Performance Indicators is not used with Secondary Option X12:

Multiparty collaboration. However, some of the secondary options and individual main options are mutually exclusive; this is not for the sake of complexity, but rather a direct response to the practicalities of project management. The logic behind the prohibition of any incompatible options becomes clear in looking at the individual secondary options.

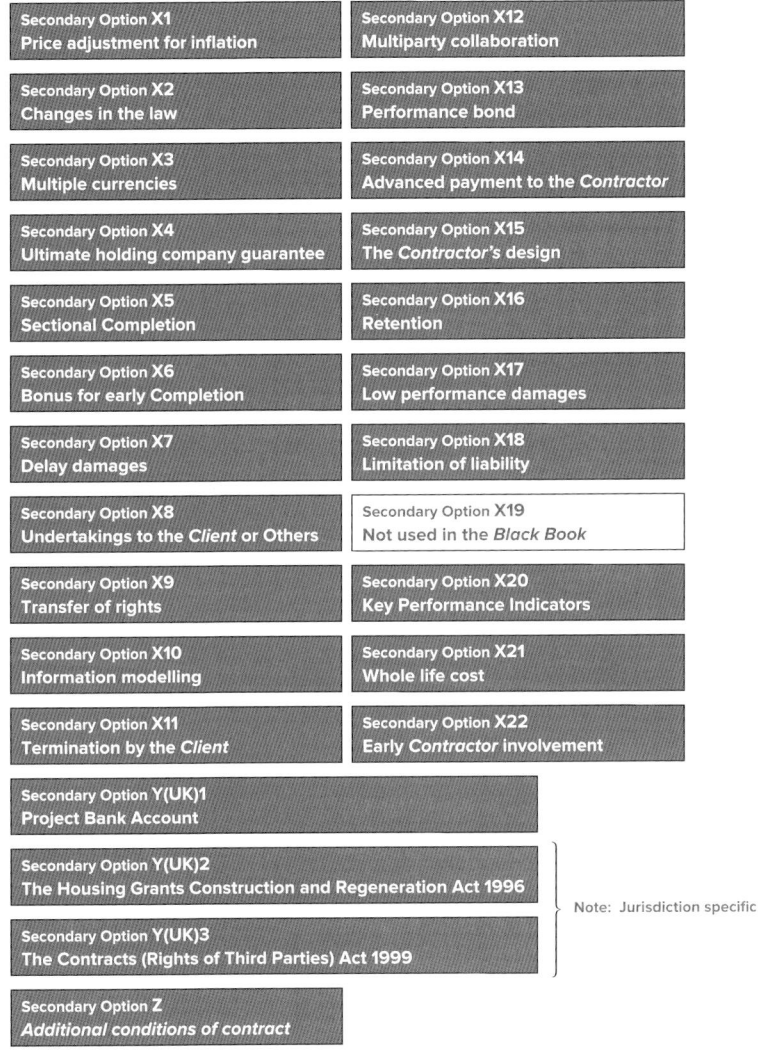

Figure 12: NEC4 Secondary Option Clauses

Secondary Option X1: Price adjustment for inflation

[Used only with Main Options A, B, C or D]

Architects and their building clients will be familiar with standard form contracts offering the potential for the Client to carry the risk of price increases. However, building clients in the UK have been reluctant for decades to voluntarily take on this risk and it is unlikely that Secondary Option X1 will be put forward on a building project in the UK. The provision is offered to cater for requirements in sectors where it is more usual for the Client to be well placed to carry the risk of price increases. It is also offered in recognition of NEC4 potentially being used in countries with high inflation, where the Contractor may be unwilling, or unable, to carry such a risk without being priced out of contention.

The reason that Secondary Option X1 is only used with Main Options A, B, C or D is simply that the payment mechanisms under Main Options E and F already entitle the Contractor to be paid on the basis of costs which include any inflation at the time of payment.

Secondary Option X2: Changes in the law

The risk of legislative changes during the currency of a contract is clearly higher under some jurisdictions than others. There is also the issue of whether a change in the law in a particular country is applied retrospectively or not. The impact on construction contracts will consequently be quite variable, depending on where the project is. Looking specifically at building contracts in the UK, the perceived risk is very low, although the procurement route will have an influence. For example, the risk of changes in the Building Regulations per se is quite high; however, because such changes do not apply retrospectively, and adequate notice is given of their coming into force, the actual risk on a particular project is normally quite low. The procurement route will have an impact on the risk as it may make a difference as to which party to the building contract would be liable for post-contract legislative changes. Using the same example of Building Regulations, the Contractor might be at greater risk under a design and build procurement route, where the time period is long between entering into contract and sign-off of design packages requiring Building Regulations approval.

Architects and their building clients will need to consider this secondary option on its merits for a particular project. Typically, it is not an option which would be considered necessary on many building projects in the UK. Where the Contractor is to be responsible for large amounts of design, or is working on a project of long duration, there might be a financial advantage to the Client at tender stage in incorporating this secondary option and accepting a risk which would otherwise rest with the Contractor.

In the case of working on projects in other countries, architects would need to be familiar with the differences in default risk in those countries. For example, the risk of changes in the law pertaining to technological standards as the project progresses, i.e. not retrospectively, would as a default rest with contractors in Germany.[87]

---

87   anerkannte Regeln der Technik [recognised rules of technology (state of the art)].

Secondary Option X3: Multiple currencies

[Used only with Main Options A or B]

It is increasingly common on construction projects for components to be imported from other countries. NEC4 is designed to be used potentially in any country, so whether, for example, Japanese air handling units and German curtain walling are being imported for a UK building project, or Swiss precast concrete elements are being exported for a Chinese hydroelectric project, there are many instances where contractors may have to pay for components in more than one currency.

The idea behind Secondary Option X3 is that the Client may be in a better position to carry the risk of currency fluctuations than the Contractor. In such a case, by incorporating Option X3, the Client can agree to pay for certain items in their currency of origin (e.g. yen or euros), while other items are paid for in the default currency of the contract[88] (e.g. pounds sterling). The use of Secondary Option X3 therefore allows the exchange rate risk to be managed by the Client rather than the Contractor.

The amount of payment in other currencies is capped in the contract to an agreed level and an exchange rate is stated. In practice, this means that where Secondary Option X3 is used, the Client can manage the exchange rate risk in quite a sophisticated manner, such as by the use of hedging finance with forward contracts, etc. This facility is particularly critical in times of exchange rate volatility, for example, the Brexit period.

Clients and their architects may feel that this Secondary Option X3 requires excessive financial management skills; however, it should be seen as the option that it is, and it is clearly more suited to certain clients than others. It should also be taken into consideration that some contractors are perfectly able to manage exchange rate risk without detriment to clients – an obvious example of this would be a multinational contracting firm with offices in both countries and therefore constantly using both currencies.

Any decision to use Secondary Option X3 will be entirely project- and organisation-specific.

Secondary Option X4: Ultimate holding company guarantee

This will be a familiar concept to architects who are used to working on larger projects. Experienced clients who are building high-value projects often require some form of security for the performance of contractors. A parent company or ultimate holding company guarantee may be an appropriate form of security, albeit it relies on a particular contractor having an ultimate holding company.

This may be a useful provision on large-scale building projects and it is relatively easy to implement by virtue of it being a standard optional clause.

---

88   Stated in Contract Data Part One.

Where Secondary Option X4 is being considered for a building contract, it should ideally be done in the context of considering as an alternative Secondary Option X13: Performance bond. While the two secondary options are not mutually exclusive, it might be considered somewhat extreme to include both; it might also be unnecessarily expensive, given that both forms of security will normally attract a fee or premium payment.

Secondary Option X5: Sectional Completion

This will be a very familiar concept to most architects and, as in other standard form contracts, is a provision which allows different parts of a project to be completed sequentially. It is intended for use where a client requirement for such sequential completion is foreseeable from the outset.

Secondary Option X5 can be implemented on the basis that the entire project is split into particular sections, or on the basis that critical parts are identified as discrete sections, with the remainder being left within the 'whole of the works'. When using Secondary Option X5, ensure that there is complete clarity as to which parts of the works fall into a particular section; depending on the nature of work to be done, this may require detailed written and/or graphic description to be included within the Scope.

The Contract Data needs to be filled out with particular care when Secondary Option X5 is used. If the entire project has been split up into sections, then it follows that in the Contract Data the *completion date* for the whole of the works will be the same date as the *completion date* for the last identified section. If only certain critical parts have been identified as discrete sections, then the *completion date* for the whole of the works will be a later date than the *completion date* for the latest identified section.

In using Secondary Option X5, it follows that while not all of the works need be allocated to one of the sections, if a project team feels more comfortable doing so, then the last identified section would effectively be on an 'and everything else' basis, and the *completion date* for the whole of the works would be the same date as the *completion date* for that last identified section.

Secondary Option X6: Bonus for early Completion

This is likely to be an unfamiliar concept to most project teams in the building sector. It is perhaps helpful to see the idea of clients rewarding contractors for finishing early as no more than the converse of clients penalising contractors for finishing late. That is not to say that there is any necessary connection between the two under any form of building contract, but rather a way of comprehending the partnering principles behind Secondary Option X6, i.e. incentivisation and reciprocity of potential benefits.

In completing the Contract Data for Secondary Option X6, note will need to be taken of whether Option X6 is to operate on a single *completion date* or on Sectional Completion dates because Secondary Option X5 is also included.

Secondary Option X7: Delay damages

This will be a known concept for most project teams in the building sector. As in other standard form contracts, this is a provision which allows clients to legitimately penalise contractors under the contract for finishing late. In implementing Secondary Option X7 under English law, it is important to remember that actual penalties are unenforceable and that the delay damages entered into the Contract Data must remain a genuine pre-estimate of the Client's loss if the Contractor were to finish late.

Architects using NEC4 for the first time may be somewhat surprised to find that delay damages are optional, as similar damages are a well-established norm under other standard form contracts.[89] However, experienced NEC users are likely to look at the range of contractual measures available to incentivise contractors to stay on programme. It would be a waste of NEC4's potential to only use Secondary Option X7 because of its familiarity. In isolation, a negative incentive is still a somewhat blunt instrument with which to tackle the issue of keeping complex building projects on programme. Hence, there is an argument for at least considering the relationship between Secondary Options X7 and X6.

There is also a need to look at an NEC4 contract holistically; the perceived need to include Secondary Option X7 would almost certainly be judged differently on a building project under Main Option A than under Main Option E.

As with Secondary Option X6, in completing the Contract Data for Secondary Option X7, note will need to be taken of whether Option X7 is to operate on a single *completion date*, or on sectional completion dates because Secondary Option X5 is also included.

Secondary Option X8: Undertakings to the Client or Others

This secondary option is new to NEC4 and provides a standard mechanism for giving parallel rights to specific parties. Architects will be accustomed to the requirement on some building projects for collateral warranties to be obtained, whether, for example, from subcontractors to clients, or from contractors to funders. Secondary Option X8 provides a clear way of incorporating such requirements into an NEC4 contract by reference to the Contract Data and the Scope, thus ensuring any required undertakings are obtained in the required format and in a known timescale.

Secondary Option X9: Transfer of rights

This secondary option is new to NEC4 and provides for design copyright to be transferred to the Client from the Contractor and from subcontractors in accordance with requirements stated in the Scope. Whether architects will consider this secondary option to be helpful will depend on the project and the procurement route; however, as with all the secondary options, it should be discussed with building clients in individual cases.

---

89 Liquidated and ascertained damages.

Secondary Option X10: Information modelling

Secondary Option X10 is new to NEC4 and provides detailed procedures for the incorporation of information modelling into projects. Architects working with building clients will be used to referring to BIM;[90] however, NEC4 uses the generic description because of its multidisciplinary application. Depending on the nature and size of projects, many architects and their building clients will regard BIM procedures as essential rather than optional, in which case they should incorporate Secondary Option X10 in their NEC4 contracts.

Secondary Option X11: Termination by the Client

This provision for the Client to terminate without specific reason used to be part of Core Clause Section 9 in NEC3, but it has been taken into the secondary option level in NEC4. This change offers greater reciprocity of rights and obligations between the parties. Architects advising building clients would be well advised to draw this change to their clients' attention and discuss whether the inclusion of Secondary Option X11 might be desirable on a particular project.

Secondary Option X12: Multiparty collaboration

[Not used in conjunction with Secondary Option X20]

Secondary Option X12[91] is a subject in its own right and is discussed in more detail in the context of collaborative working as a whole – see Chapter 4.

The key message is that Secondary Option X12 acts as extra 'cement' with which to ensure the partnering relationship stays together and can achieve sophisticated aims; it is not the sole vehicle for creating a partnering relationship in the first instance: that is NEC4 itself.

Secondary Option X13: Performance bond

As with Secondary Option X4, providing a form of security to clients for the performance of contractors will be a familiar concept to architects used to working on larger projects. In this case, the security is ultimately provided by a bank or insurer not by the contracting organisation.

While this provision is in principle relatively easy to implement, being a standard optional clause, care must be taken over stating the amount of the performance bond required,[92] describing the form of the bond required[93] and outlining the type of surety[94] which will be acceptable.

---

90 Building Information Modelling.
91 See Chapter 4, *Partnering*.
92 To be stated in Contract Data Part One.
93 To be stated in the Scope.
94 The Project Manager has to accept the bank or insurer providing the Secondary Option X13: Performance bond.

Where Secondary Option X13 is being considered for a building contract, sensibly it would not be in addition to Secondary Option X4: Ultimate holding company guarantee, as this would normally be considered somewhat excessive. This is not an NEC4-specific point, but rather common sense in the context of how best to protect clients against potential non-performance of contractors.

Secondary Option X14: Advanced payment to the Contractor

This provision for the Client to make an advanced payment to the Contractor is something which will rarely be considered on some types of building project but will be regarded as almost essential on other types of building project. As with many of the secondary options, this is not a novel concept; NEC4 is merely recognising common contractual requirements and pre-drafting provisions for such requirements so that they may be picked off the contractual 'shelf' and suitably mixed with the other contract clauses to achieve the required overall 'recipe'.

An example of this secondary option being beneficial on a building project would be where a large amount of prefabricated structural steelwork was part of the Scope of the project, its lead-in time was critical, and the fabricators required the raw steel to be purchased in advance of the fabrication.

Secondary Option X14 can operate in two ways. First, the Client can simply agree to pay a sum of money in advance to the Contractor for a particular expenditure, and this sum is then paid back in instalments[95] at payment assessment intervals. Second, where large sums of money might represent an unacceptable risk, the Client may also require an advanced payment bond. In the latter case, additional attention must be paid in completing the Contract Data to include a description of the form of the advanced payment bond required[96] and the type of surety[97] which will be acceptable.

Secondary Option X15: The Contractor's design

This secondary option is a significant one for architects, especially those who regularly work for design and build contractors.

Essentially, the position on design liability under English law is different from other jurisdictions, in that 'reasonable skill and care' is an obligation peculiar to common law. The default under most jurisdictions is 'fitness for purpose' and because NEC4 is designed to be operable under any jurisdiction, an active choice has to be made to limit liability to 'reasonable skill and care'. If the Scope is drafted in clear enough terms, that limitation of liability can legitimately be made in the Scope. However, by incorporating Secondary Option X15, the standard of care can be globally limited, which in many cases may be preferable.

---

95 Stipulated in the Contract Data.
96 To be stated in the Scope.
97 The Project Manager has to accept the bank or insurer providing the Secondary Option X14: Advanced payment to the Contractor.

Certainly, an architect contemplating engagement within a design and build procurement route by contractors who are themselves under an NEC4 main contract would do well to ask whether Secondary Option X15 has been incorporated!

Secondary Option X16: Retention

[Not used with Main Option F]

It may come as a slight surprise to architects that retention is an optional concept under NEC4. It behoves us perhaps to reconsider the intended purpose of retention, in that there are apparent instances of it being abused more than used under some standard form contracts.

Retention is generally accepted to serve a useful function in incentivising contractors to return to site to deal with the eventuality of any defects which manifest themselves post-completion. Under some standard form contracts, it is also often used as leverage to ensure contractors continue to work on elements of a building which were manifestly incomplete, or defective, when completion was certified in order to give access to a Client. In either scenario, retention that is deducted from the outset of a project is a fairly blunt instrument with which to control post-completion quality and in any case tends to push contractors towards negative cashflow.

In contemplating NEC4, there is a need to look at retention in the context of collaboration and incentivisation. Returning to the carrot and stick analogy, it is arguable that retention has always been used as a 'stick', not a 'carrot'. Not only are sticks inherently dangerous in the arena of collaboration but also using a stick from the outset of the Contract to tackle a potential post-completion issue seems perverse. NEC4 offers two alternative approaches. First, the approach of deciding that other contractual provisions could provide adequate protection from potential defects and therefore Secondary Option X16: Retention will not be necessary on a particular project. Second, where it is decided that retention is necessary in principle, Secondary Option X16 envisages that it will not be deducted from the outset but only from a point when the cumulative value of the works has reached a certain value, i.e. the so-called 'retention free amount'.[98] Only after this value has been reached, which typically would be shortly before completion, will retention be deducted; in practice this significantly reduces the period during which the Contractor has to finance the retention sum, without reducing the post-completion protection to the Client.

This secondary option is not used with Main Option F on the policy principle that retention is inappropriate to projects with predominantly subcontracted work.

---

98   To be stated in the Contract Data.

Secondary Option X17: Low performance damages

This particular secondary option is unlikely to apply in the context of a building project. It is primarily intended to provide protection to clients on projects where the scope of the intended work encompasses items, or installations, when it may not be a simple case of either they work or they do not work, but rather that they do not work quite as well as they should! In practice, Secondary Option X17 will provide useful protection to clients on projects such as industrial/processing plants where the designed output may not be fully achieved, either temporarily or permanently.

This secondary option would also protect the Contractor on certain types of project: a defect causing low performance could be accepted under the terms of a contract incorporating Secondary Option X17, rather than having to be treated as a breach of contract without Secondary Option X17.

Secondary Option X18: Limitation of liability

This secondary option was introduced with the third edition of NEC. It was probably a policy decision based on consumer demand to be able to cap liability at some point, irrespective of the type of project and contract value.

Certainly, on most building projects being carried out under English law, there has been a perceptible trend to limit liability in some manner. The reason this is a secondary option under NEC4, rather than being within the core clauses, is that the contract flexibility is of paramount importance and it is not desirable under any jurisdiction to have core clauses which might get struck out. Secondary Option X18 is certainly one of the secondary options which architects should discuss carefully with their clients before any decision is made regarding its incorporation or otherwise. In the case of contractor negotiations, it is prudent to consider discussing the inclusion of Secondary Option X18 with the potential contractors.

In the context of a design and build procurement route under an NEC4 main contract, an architect employed by the Contractor would sensibly wish to see Secondary Option X18 incorporated because liabilities should relate to insurance cover[99] and no architect is going to carry professional indemnity insurance with limitless cover.[100]

---

99   Secondary Option Clause X18.3.
100 Note: Option X19 is not used (reserved for TSC – Term Service Contract).

Secondary Option X20: Key Performance Indicators

[Not used in conjunction with Secondary Option X12]

The only reason Secondary Option X20 should not be used with Secondary Option X12 is because the latter potentially offers an even more sophisticated approach to Key Performance Indicators, linked into incentivisation; these secondary options are therefore designed to be used on an either/or basis, not simultaneously.

Key Performance Indicators, or KPIs, as they are commonly known, became a feature of many contracts where a collaborative ethos was being encouraged. Their role is essentially to assist in benchmarking relative performance in specific areas within a construction project and to incentivise improved performance. Secondary Option X20 is intended to provide a relatively simple contractual basis upon which performance targets can be stated, measured and monitored.

Secondary Option X21: Whole life cost

This secondary option is new to NEC4 and is an opportunity for the Client to benefit from the Contractor proposing a change to the Scope to reduce the costs associated with operating and maintaining a project.

Architects are likely to be considering whole life cost in any case and may deem this option unnecessary if they have been involved in the design of a building project from the outset. However, if architects were employed at a later stage under a design and build procurement strategy, they might consider that operation and maintenance costs could be reduced, and they would be well placed to assist the Contractor in preparing the quotation required under Secondary Option X21.

Secondary Option X22: Early Contractor involvement

[Used only with Main Options C or E]

Secondary Option X22 is new to NEC4 and provides for the Contractor being involved from an early stage of a project and contributing to the control of a budget.[101] Architects will need to consider carefully how this secondary option relates to the relative expertise of all project contributors, and the proposed allocation of design responsibility, in order to decide whether it is likely to be helpful for a particular project or client.

Secondary Option Y(UK)1: Project Bank Account

This UK-specific secondary option is new to NEC4 and provides for a bank account to be opened by the Contractor which will enable security of payment from Client to Contractor and from Contractor to Named Suppliers.

---

101 Secondary Option Clause X22.1.

A trust deed is executed between the Client, Contractor and Named Suppliers. A joining deed is envisaged for additional suppliers to be included at a later date, after the original trust deed has been executed.

This provision may be helpful on larger, higher risk projects, and architects can discuss the possibility with their clients.

It is unclear why this new provision has been classified as a UK-specific secondary option because the principle of security of payment through escrow accounts is well known under other jurisdictions. Indeed, the desire to increase security of payment might be greater in the case of cross-border contractual relationships which could be under jurisdictions other than English law.

Secondary Option Y(UK)2:
The Housing Grants, Construction and Regeneration Act 1996

[Only to be used on UK projects where the Act applies]

Secondary Option Y(UK)2 relates specifically to legislation which extends to England and Wales and where the relevant part of that legislation also extends to Scotland.[102] The purpose of Y(UK)2 under NEC4 is to ensure that the payment provisions of the legislation are complied with. (NEC4 does not deal with adjudication as part of Y(UK)2.)[103] It therefore follows that a contract for any project in England, Wales or Scotland, which comes under the definition of a 'construction contract',[104] should have Secondary Option Y(UK)2 incorporated.

Irrespective of the specific use of NEC4, architects advising clients will need to be familiar with the project types which will be deemed exceptions[105] to the definition of a construction contract, and where the payment provisions of the legislation therefore need not be complied with; i.e. in the case of NEC4, the projects which do not require Secondary Option Y(UK)2 to be incorporated.

Secondary Option Y(UK)3:
The Contracts (Rights of Third Parties) Act 1999

[Only to be used on UK projects where required]

Most UK-based architects will be familiar with the common advice to exclude the operation of the Contracts (Rights of Third Parties) Act 1999 because the legislation may be contracted out of.

---

102 Part II of the Housing Grants, Construction and Regeneration Act 1996.
103 See *Dispute resolution options*.
104 Section 104, Part II of the Housing Grants, Construction and Regeneration Act 1996.
105 Sections 105, 106 & 107, Part II of the Housing Grants, Construction and Regeneration Act 1996 and the Construction Contracts (England and Wales) Exclusion Order 1998.

Secondary Option Y(UK)3 can be seen as a slightly more subtle approach. The legislation can be contracted out of entirely by incorporating Secondary Option Y(UK)3 and by remaining silent as to any third-party identity. However, by naming a particular third party, or class of third party, in the Contract Data, it is possible to allow the Act to apply in a specifically controlled manner. In practice, this may be a preferable approach on building projects where third parties, such as funders or tenants, require rights, rather than having a number of specially drafted collateral warranties.

NEC4 per se does not alter the decision-making process in relation to whether or not to allow either collateral warranties or invocation of the Act; it is merely that Y(UK)3 facilitates the practical process where a decision has been made to either completely exclude or specifically invoke the Act.

Secondary Option Z: Additional conditions of contract

Any so-called 'Z-Clause' should be very carefully considered and if strictly necessary, it should be drafted in a style compatible with the drafting of NEC4.

The status of a Z-Clause is that it *augments* the other contract clauses; it should not purport to *amend* them.

There is a strong argument for Z-Clauses to be used only if genuinely necessary. In the early days of NEC contracts, there were many examples of copious amounts of Z-Clause drafting, much of which was superfluous and some of which was downright dangerous: it either confused the meaning of core clauses or sought to change that meaning without necessarily dealing with the consequences of an isolated change. Such overzealous drafting thankfully appears to have diminished significantly, no doubt as a result of the now widespread use of NEC contracts and an ever-improving understanding of them.

Nevertheless, architects should remain vigilant for unnecessary Z-Clauses, which could creep in as a result of clients seeking legal advice from lawyers not entirely familiar with NEC. For reasons of democracy and moving projects forward, there might conceivably be times where letting a superfluous Z-Clause remain is a reasonable course of action, in that it is very unlikely to be harmful. However, if an architect or their client is unlucky enough to be presented with a long list of Z-clauses which purport to amend core clauses, they must resist! A good clue when checking for such clauses is if the drafting style of Z-Clauses appears to be old fashioned in comparison with that of the NEC4 standard clauses. This would suggest that a lawyer, quantity surveyor or even an ill-advised architect has returned to the older style standard form contract drafting originating from the Victorian era, possibly having found the absence of such drafting in NEC4 somewhat unnerving.

This may be seen as a controversial point; nonetheless, new methods in any sphere can be controversial, and reading this book should help form an unequivocal view about whether NEC4 is a desirable step forward. Given that society is made up of some people who embrace change and some who resist change, a little self-analysis may be helpful. Architects' training is certainly based on the principle of being competent in designing and managing change, so perhaps architects are particularly well placed to assess the pros and cons of NEC4 itself and the desirability, or otherwise, of Z-Clauses for an individual building contract.

Options W1, W2 and W3: Dispute resolution options

The evolution of the dispute resolution option clauses must be seen in the light of the history of adjudication in the UK relative to the drafting of the NEC contract. Adjudication as an idea has been around for some time and, in its earliest guise, was never intended to do more than put roughly the right amount of money in the right person's pocket at roughly the right time. NEC incorporated adjudication into its Core Clause Section 9 at the outset, and it was intended to operate as a contractual procedure where either party, or both parties, felt that their own ability to resolve a disagreement amicably was overstretched. The intention under NEC contracts has always been for the adjudicator to be named in advance, normally available to act as and when needed. In the early days of NEC, it is no exaggeration to state that the adjudicator was seen simply as an extension to the parties acting in 'a spirit of mutual trust and co-operation'. It was foreseeable that the parties might hit a stumbling block in the operation of the contract and might disagree on the appropriate course of action; the adjudicator was there to assist in such a situation. In view of the adjudicator's early involvement by being named in the contract, there was a reasonable expectation that they would be familiar in general terms with the project and the contract, and would be able to make a judgement on an individual disagreement relatively quickly and easily.

It has been said retrospectively that the Construction Act, which came into force on 1 May 1998,[106] hijacked adjudication as the NEC knew it. This perception stems from the formality that statutory adjudication has engendered, including very often legal representation of the contracting parties. Such formality is in stark contrast to the initial NEC intentions, which were simply pragmatic and provided the contracting parties with impartial assistance in resolving a dispute that they had not managed to resolve unaided.

---

106 The Housing Grants, Construction and Regeneration Act 1996; Part II was brought into force by the Scheme for Construction Contracts (England and Wales) Regulations 1998.

All the standard form drafting bodies were apparently somewhat challenged by the introduction of this legislation as it was a relatively rare example under English law of the phenomenon of legislation interfering with the principle of 'freedom of contract'. This phenomenon has increased somewhat since;[107] however, the Construction Act remains surprising to some construction professionals and their clients in that it is mandatory, i.e. it cannot be contracted out of. There was considerable debate, both at the time and subsequently, over the principle of Act compliance and the need to amend standard form contracts to become 'Act-compliant' to avoid the Scheme being operated by default.

Opinions were divided over the extent to which legislation should drive standard form drafting. NEC was at the time unusual among the standard form drafting bodies in leaving the Scheme to apply to projects which came under the Construction Act,[108] while Core Clause Section 9 remained the norm for other adjudications. This original decision should be seen in the context of the international operation of NEC contracts and not wishing to include extraneous drafting in the core clause sections. Interestingly for architects, deliberately having the Scheme apply as a default, rather than maintaining Act-compliant drafting, was subsequently embraced by other standard form building contracts.[109]

The post-Construction Act status quo was maintained under the NEC second edition for a short time, until the first official legal comment on the NEC contract, specifically the NEC second edition Secondary Option Y(UK)2.[110] The matter revolved around whether the requirement for the issue of a Notice of Dissatisfaction as a precondition for a dispute being deemed to have arisen a minimum of four weeks later constituted a breach of the right conferred under the Construction Act to commence an adjudication 'at any time'.

This case engendered a standard form drafting body debate on a dispute resolution clause with similar drafting to Y(UK)2,[111] and the consequence when the review of the NEC second edition provisions took place was a decision to move all adjudication provisions outside both Core Clause Section 9 and the secondary option clauses. The Construction Act indirectly caused the drafting of a new breed of NEC option clauses, the so-called 'W' options for dispute resolution, which have been carried forward from the third edition into NEC4, and Options W1 and W2 have been augmented with Option W3.

Under NEC4, one of the three following dispute resolution options must be chosen:

Option W1

Option W1 is the dispute resolution procedure used unless the United Kingdom Housing Grants, Construction and Regeneration Act 1996 applies. This remains a contractual adjudication, using contractual adjudication procedures.

---

107 E.g. The Contracts (Rights of Third Parties) Act 1999.

108 NEC second edition Secondary Option Y(UK)2: The Housing Grants, Construction and Regeneration Act 1996.

109 E.g. JCT Standard Forms of Building Contract.

110 Judge Toulmin's obiter comments in: *John Mowlem & Company PLC v. Hydra-Tight & Company PLC (t/a Hevilifts)* 2000.

111 ICE Conditions of Contract, seventh edition, September 1999.

Option W2

Option W2 is the dispute resolution procedure used in the United Kingdom when the Housing Grants, Construction and Regeneration Act 1996 applies. This enables a statutory adjudication, using contractual adjudication procedures in place of the adjudication procedures in the Scheme.

The intention has been to make the language and terminology in Options W1 and W2 as similar as possible while ensuring that Option W2 is Act-compliant. The crucial difference between the two options is that in its Adjudication Table Option W1 still only allows the contracting parties to refer a dispute to the adjudicator after specific notification periods between the contracting parties, whereas Option W2 must allow either contracting party to refer a dispute to the adjudicator *at any time* to remain Act-compliant.

NEC4 introduced the requirement under Option W1 and the opportunity by agreement between the parties under Option W2 for the parties to attempt to resolve a dispute amicably by referring it initially to their Senior Representatives, identified in the Contract Data. The Senior Representatives have up to three weeks to resolve the dispute by any method they deem appropriate before any remaining areas of dispute will be referred afresh to adjudication.

Option W3

Option W3 is the dispute resolution procedure used when a Dispute Avoidance Board has been set up and when the United Kingdom Housing Grants, Construction and Regeneration Act 1996 does not apply.

This allows a Dispute Avoidance Board of one or three members to be appointed at the outset (under the NEC4 Dispute Resolution Service Contract) and to know the details of the project prior to any potential dispute. In the event of a potential dispute arising, the Board makes a recommendation intended to avoid it developing and neither party may refer a dispute to the *tribunal*[112] unless the Board has acted first.

Architects are unlikely to come across Dispute Boards except on larger, multidisciplinary and possibly multinational projects; however, it is important that the potential benefits of Option W3 are considered on projects where the Construction Act does not apply.

---

112   Whether arbitration or litigation, members of the Dispute Avoidance Board cannot be called as witnesses.

Contract Data

Role of the Contract Data

The Contract Data is a sophisticated document defining the project-specific parameters clearly in relation to the generic contract conditions, i.e. the core clauses. The requirement for both Client and Contractor to prepare Contract Data (Part One and Part Two respectively) ensures parity from the outset and is consistent with the universal reciprocity between parties to any of the NEC family contracts.

Part One: Data provided by the Client

The principle of project-specific parameters relative to generic contract conditions will be familiar to architects from other standard form building contracts. The dual role of such project-specific parameters is also well established, i.e. the initial function to define requirements at tender stage and the subsequent function to make the contract operable at construction stage. NEC4 does not deviate from these principles and most architects should find Contract Data Part One self-explanatory.

An area for caution is where the additional items are completed for the chosen main option and for any chosen secondary options, as these are just as important as the entries against the core clause sections. Also, if a large number of secondary options, or if certain option combinations, are chosen, particular care is required to complete the Contract Data Part One correctly.

Part Two: Data provided by the Contractor

The reciprocal requirement for contractors to provide project-specific information at tender and/or negotiation stage will be less familiar to architects than Contract Data Part One. However, it is easy to see how this is intended to operate and give much greater control over managing projects subsequently.

The key point for architects to note is that in the context of assessing compensation events, the information completed in Contract Data Part Two ensures both objective and foreseeable assessment bases, as well as being contractually binding. In practice, this is likely to make a major contribution to dispute avoidance on many building projects.

Schedule of Cost Components

Role of the Schedule

Possibly the most important point to emphasise about the schedule is that it is not any sort of price list. The word 'component' is key, in that the schedule provides a comprehensive method for breaking down costs into recognisable and objective components.

It takes a little practice to fully understand the principles of the schedule, but it will become apparent to architects that the objectivity it is based upon is a major step forward compared with assessing costs in a subjective manner.[113]

Schedule of Cost Components

This 'full' version of the schedule applies with Main Options C, D or E.

Short Schedule of Cost Components

This short version of the schedule always applies to Main Options A and B, and architects may therefore find that it is worth familiarising themselves with this version first.

Neither schedule applies to Main Option F.

Both schedules split cost into the following component categories:

- People
- Equipment
- Plant and materials
- Subcontractors
- Charges
- Manufacture and fabrication
- Design
- Insurance.

It is the extent of subdivision within each category which is greater under the full schedule than under the short schedule.

Defined Cost

As the name implies, cost is not necessarily exactly what is spent, but rather sums which are defined as being entitlements to payment. The definition itself takes place relative to the Schedule of Cost Components and Defined Cost will therefore entitle payment of sums falling under the component categories.

Defined Cost record-keeping requirements vary under the six main options and clearly, entitlement to payment relies on the required records being available.

---

113   E.g. on a 'fair and reasonable' basis.

Disallowed Cost

The corollary to payment entitlement depending on Defined Cost as recorded under the Schedule of Cost Component categories is that where a cost falls outside those categories it will be disallowed, i.e. Disallowed Cost. Costs will also be designated as Disallowed Cost where records do not justify those costs and in circumstances where contractual procedures have not been followed properly, for example, the Scope has not been followed or required communications have not taken place. The rules as to when a cost might be classified as Disallowed Cost are to be found in the specific definition of Disallowed Cost given under Main Options C, D, E and F.[114]

Priced contracts under Main Options A and B do not recognise Disallowed Cost, because the mechanism for changing the prices is controlled solely by the compensation event procedure.

Scope

Scope used to be called 'Works Information' under previous NEC editions, but it has been renamed in NEC4 to create common terminology with NEC professional services contracts, which have always referred to 'Scope'. The meaning of Scope can be summarised as the entire information a Contractor requires in order to know what to build (and how). The contract definition remains as follows:[115]

Scope is information which either:
- specifies and describes the *works* or
- states any constraints on how the *Contractor* provides the Works.

and is either:
- in the documents which the Contract Data states it is in or
- an instruction given in accordance with this contract.

The quality of the design documentation is of paramount importance under NEC4. Some designers are concerned that there is nowhere to hide, although arguably there never should have been. (Even older standard contract forms in their later versions have introduced greater accountability for information release for construction.)

The NEC does not distinguish between tender documents and contract documents, but instead gives the generic term of Scope to all design documentation, whether it be drawings, descriptive or performance specifications, or schedules.

Architects should ideally treat the designation Scope simply as a neat receptacle (whether bucket or cardboard box!) in which to meticulously place *all* production information – preferably once and once only, and in the right format and order for ease of comprehension.

---

114 Clauses C11.2(26), D11.2(26), E11.2(26) & F11.2(27).
115 Core Clause 11.2 (16).

That is not a definition, as it is not an exclusively NEC-related aim; however, because of the status of the Scope, NEC4 will be found less forgiving than some standard form contracts if that aim is not pursued.

For the avoidance of doubt, any architect wondering whether and/or where to place an NBS specification or Preliminaries in an NEC package should immediately reach for the Scope. Nevertheless, such documents need to have been prepared with greater rigour than may be the default norm if they are not to risk acquiring unwanted characteristics.

The Scope is a vital part of the NEC documentation and functionality. Architects need to be comfortable with the principle of it to ensure that their design information has the intended contractual status. It is also important to acknowledge the status of the Scope as a live set of documents – it can be changed during the currency of the Contract to respond to project requirements.[116]

Site Information

The NEC separates the design of the construction from the risks inherent in the context of that construction. Site Information can be summarised as all available information for the Contractor to assess the context of the construction and therefore the risks. The contract definition is as follows:[117]

> Site Information is information which:
> * describes the Site and its surroundings and
> * is in the documents which the Contract Data states it is in.

Completeness of Contractor's perception of Commercial Risk

Many commercial risks (e.g. ground conditions) are likely to be greater in the Site Information than in the Scope. A serious inaccuracy in the Site Information may lead to the Scope having to be amended, for example, foundation redesign. A Contractor relies on accurate and complete Scope information and Site Information to be in a position to accurately assess the commercial risks inherent in a particular project at tender stage. It may be advantageous for the Client to carry the risk of incomplete or inaccurate Site Information if the unknowns are such that the Contractor would be pricing an unquantifiable risk. Allocation of risk between the parties to a building contract will affect cost on a similar principle to the cost of insurance being related to the extent of cover.

---

116 See Chapter 3, *Change control*.
117 Core Clause 11.2 (18).

The Agreement

NEC4 envisages that the contract conditions are incorporated by reference and that the parties to the contract enter into an Agreement which accurately describes the required generic clauses and project-specific parameters.

It is open to the parties to an NEC4 contract to draw up their Agreement as a bespoke document. This stems from the envisaged flexibility of who will use the contract, for which project, in which country and under what jurisdiction.

As with all contracts entered into under English law, the parties may execute the Agreement as a Simple Contract[118] or expressed as a Deed.[119]

---

118   Six years' liability under The Limitation Act 1980.
119   Twelve years' liability under The Limitation Act 1980.

3 Contract machinery

The purpose of this chapter is to explore on a pragmatic basis the key areas where NEC4 differs most from old-style standard form contracts. Architects will find these areas decisive in relation to contract administration under NEC4.

People

The 'players' in NEC4 will not be entirely familiar to all new users as distinct roles have been created to facilitate flexible use of people's skills to suit individual projects and project teams. This section details the people in NEC4 and their roles.

Client and Contractor

The *Client* and the *Contractor* are the parties to a *Black Book* NEC4 building contract and are therefore the signatories, as with other construction contracts. The Agreement for signature is drawn up as a bespoke document for each NEC contract, similar to the Contract Data, to take account of project-specific aspects. The *Black Book* building contract is then administered by the Project Manager and the Supervisor.

The term *Client* is also used across the NEC4 contract family: for example, in the NEC4 Short Contract (where the complexity of the project does not warrant contract administration by a Project Manager and a Supervisor) and in the Professional Service Contract (in the context of engaging consultants).

Project Manager and Supervisor

There is a separation of the time and cost contract administration and the quality inspection contract administration function within the contract, and a potential separation of both of these functions from any pre-contract design. The Project Manager (PM) performs the time and cost contract administration function and may potentially come from any appropriate discipline. On building projects, architects should be well equipped to perform this role, although many quantity surveyors or cost consultants have acted as PM under NEC contracts. The Supervisor performs the quality inspection contract administration function, and again, on building projects, architects should be well equipped to perform this role, although other disciplines, including clerks of work, are capable of taking on the Supervisor role. The key to the Supervisor's correct contract administration will be a good understanding of the Scope and the ability to assess compliance with it when inspecting the works. Quality is to be assessed relative to the standards stipulated in the Scope documentation.

In practice, Clients are given the freedom to appoint the appropriate bodies to each of these distinct functions: (designer), PM and Supervisor. It is perfectly possible for one Consultant, who may be the Architect, to act as lead designer and, additionally, as both PM and Supervisor, if required. (The identity of the PM and the Supervisor will be stated in Contract Data Part One).

In the early days of NEC, architects tended to believe that the contract was not for them as it made no mention of the Architect. However, even those standard form contracts that do refer to the Architect are doing so in the Architect's capacity as contract administrator, not as designer, and it therefore follows that NEC is not in any way excluding architects from fulfilling a contract administration role. The roles of PM and Supervisor are deliberately not discipline specific, to remain compatible with the notion of NEC contracts themselves not being sector specific.

It is entirely logical that architects will be suitable contract administrators for building projects which are to be built under an NEC4 form of contract. The only decision which needs to be made to gain maximum advantage from the flexibility of NEC4 is whether that contract administration would benefit from the separation of the time and cost function and the quality function. In practice, that decision will be influenced both by the procurement route for a particular project and by the preferences and skills of individual architects.

One important aspect of the contract administration roles of both PM and Supervisor is the question of authority. These merit special attention in the context of architects' training and the customs developed over many years of older standard form contracts:

The authority of the PM

Architects will be very familiar with the concept of impartiality in administering traditional standard form contracts. While this no longer includes a quasi-arbitral role, it nevertheless remains well established through testing in the courts that an Architect administering a traditional standard form contract has a duty to act impartially and is 'holding the balance' between Employer and Contractor.[120] NEC4 language throughout the contract makes the PM unequivocally the *Client's* person, which has led to some debate as to what extent the PM has a duty to act fairly. In view of the relatively objective assessment criteria contained in the NEC contract drafting, there seems to have only ever been limited scope for the PM to unfairly favour the *Client*; however, this point received brief judicial consideration[121] (in the context of an amended version of the NEC second edition), and this underlined the logic that the PM should maintain an impartial stance in performing contractual duties such as assessment and certification.

---

120 *Sutcliffe v. Thackrah* 1974.
121 Mr Justice Jackson's obiter comments in *Costain Ltd & Others v. Bechtel Ltd* 2005.

The authority of the Supervisor

Architects will also be very familiar with the advice given in relation to administering traditional standard form contracts to state that the Architect only *inspects* and does not *supervise*. It follows that some architects may be reluctant to take on the role of Supervisor for fear of assuming responsibility inferred from the title. This is an example of the lack of transferability of terms between older and newer standard form contracts. The only sensible advice in relation to architects, NEC4 and the title Supervisor is 'fear not'. It is reasonable to expect the authority of the Supervisor to be construed relative to the description of the role within NEC4 and, provided an architect is comfortable with that role, there should be no grounds for concern in an architect being the Supervisor.

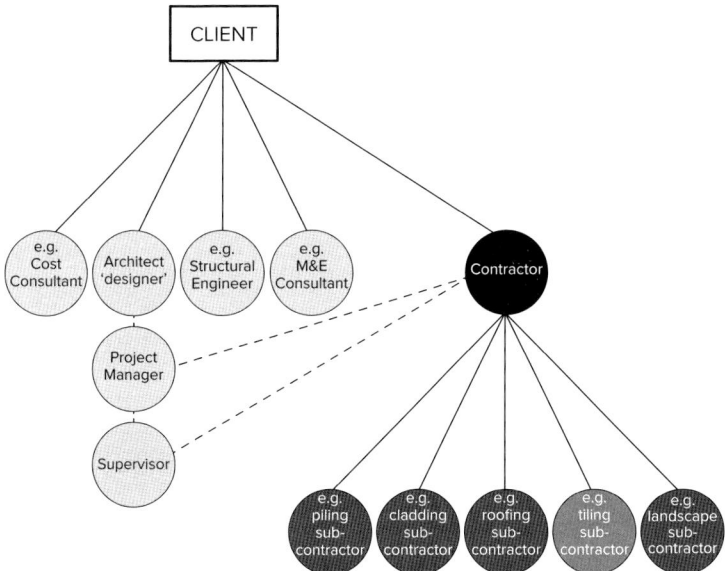

Figure 13: NEC4 'Traditional Procurement Architect Acts as PM and Supervisor

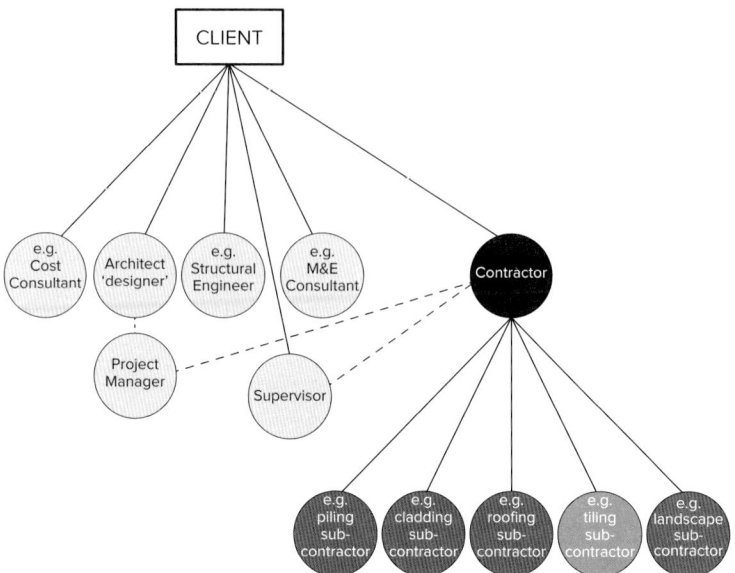

Figure 14: NEC4 'Traditional' Procurement: Architect Acts as PM

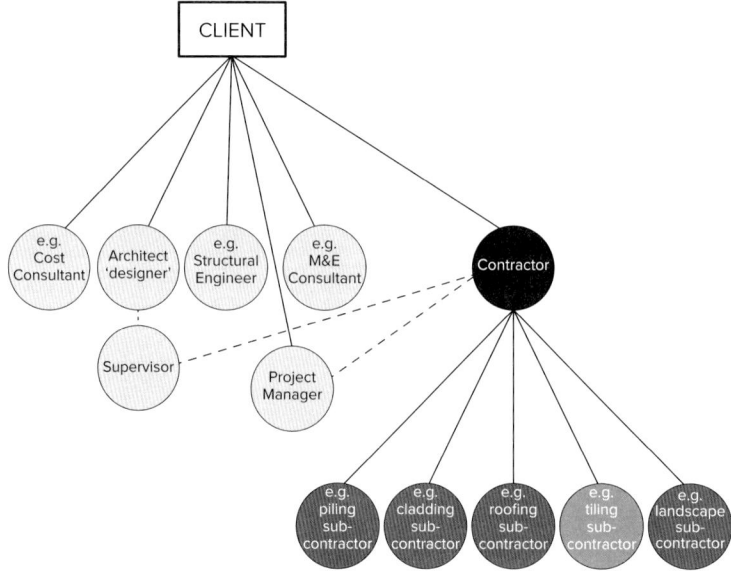

Figure 15: NEC4 'Traditional' Procurement: Architect Acts as Supervisor

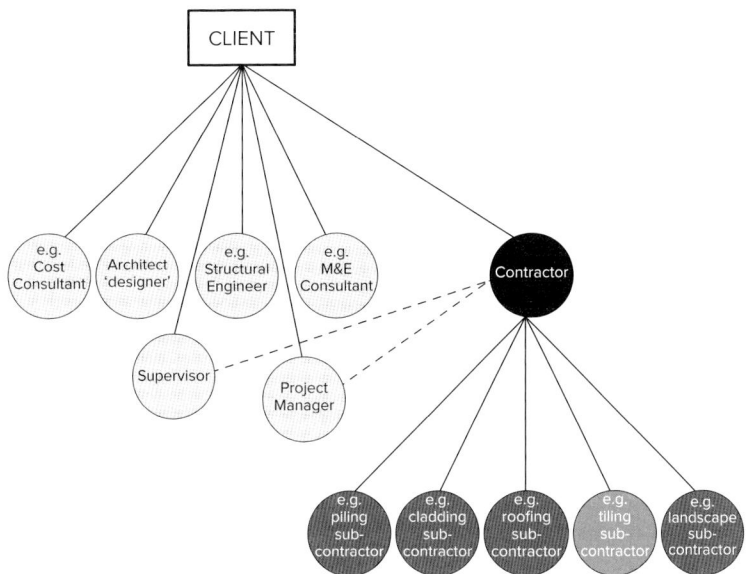

Figure 16: NEC4 'Traditional' Procurement: Architect does not administer Building Contract ('designer' role only)

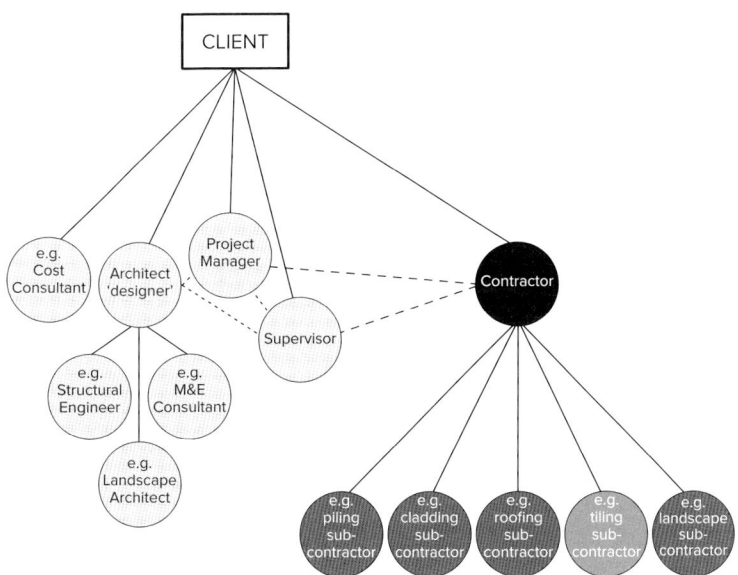

Figure 17: NEC4 'Traditional' Procurement: Architect employs Subconsultants (whether or not acting as PM/Supervisor)

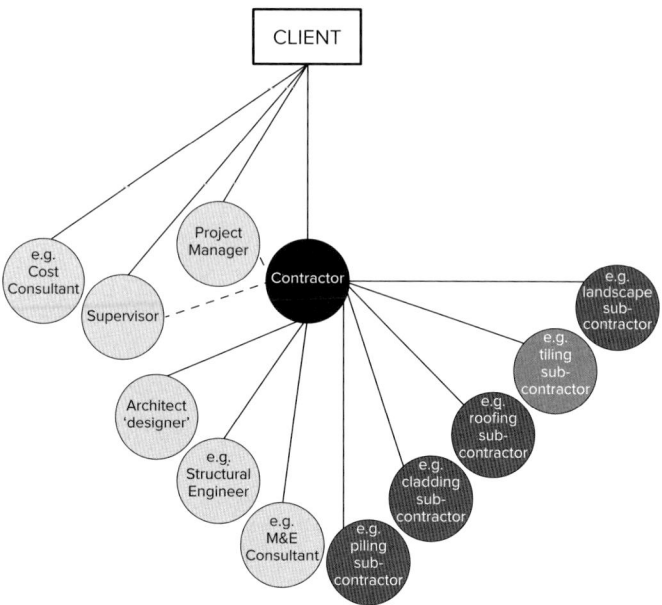

Figure 18: NEC4 'Design & Build' Procurement: Architect Employed by Contractor

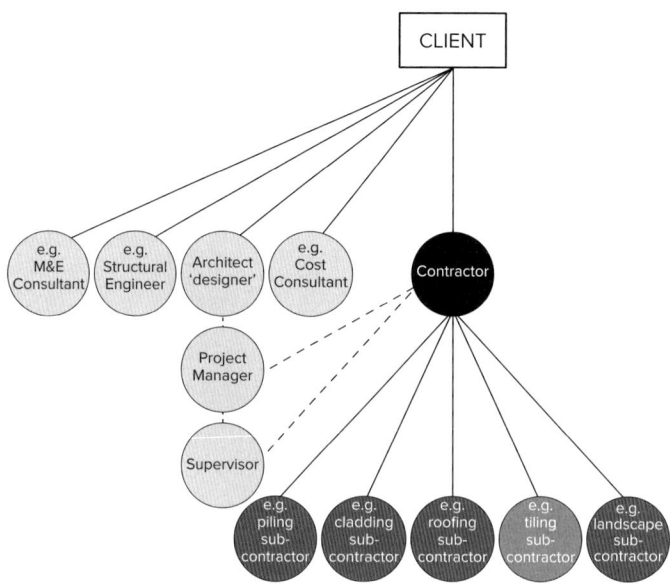

Figure 19: NEC4 'Management' Procurement: Architect Acts as PM and Supervisor

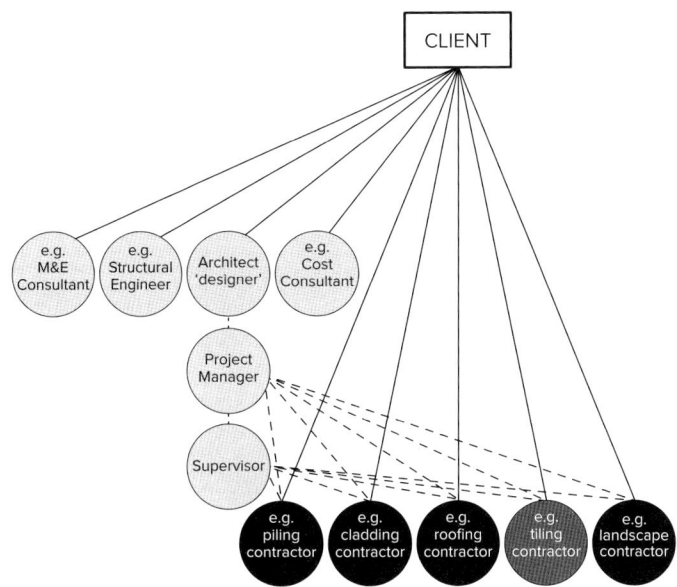

Figure 20: NEC4 'Construction Management' Procurement: Architect Acts as PM and Supervisor

Subcontractor

Subcontractors are all bodies in contract with the Contractor to provide part of the works, whether or not they carry design responsibility, and even if they are not employed under the NEC subcontract, although the PM has to accept any other conditions of subcontract.

Others

Others are defined in the *Black Book* NEC4 building contract as people or organisations who are not the Client, the Project Manager, the Supervisor, the adjudicator or a member of the Dispute Avoidance Board, the Contractor or any employee, Subcontractor or supplier of the Contractor. (This would in practice include Consultants under the PSC.)

Adjudicator

The adjudicator is the person to whom any 'crystallised' dispute during the contract will be referred in the first instance (after Core Group review where NEC Option X12 is included) and the person who is empowered to settle such a dispute on an interim binding basis. If either party is dissatisfied with the adjudicator's decision, they may take the dispute afresh to the chosen Tribunal (litigation or arbitration) after Completion. (Note the distinction between a statutory and a contractual adjudication.)

Consultant

The Consultant can be any discipline of specialist who enters into a PSC with the Client who, in this context, can be either the project sponsor or a Contractor in a design and build procurement strategy.

Promoter

The Promoter is the project sponsor in collaborative relationships where NEC4 Secondary Option X12 is included. The distinction between Promoter and Client should be noted, as many NEC4 contracts that include Secondary Option X12 may have a Client who is not the project sponsor. For example, an architect may employ a consulting engineer as a subconsultant under an NEC4 Professional Service Subcontract (PSS) which includes Secondary Option X12 and, in that instance, the Client (architect) will be distinct from the Promoter (project sponsor).

(Architects acquainted with previous editions of NEC prior to NEC4 should note that Promoter has replaced Client and that Client has replaced Employer.)

Programme

The programming requirements under NEC4 are simultaneously onerous and desirable.

The status of the NEC4 programme is that it is mandatory. The form of the NEC4 programme is really nothing short of a critical path style programme which needs to be prepared and kept current to comply with the NEC4 Core Clause Section 3 requirements. The function of the NEC4 programme is essentially to integrate the contract obligations with efficient project management.

Pricing and payment

The selected main option clearly has a fundamental impact on the manner in which NEC4 contracts are initially priced and the subsequent procedures that must be operated both to make interim payments and to adjust payments in the event of authorised changes. It is key to understand that the defined calculation method for the PWDD[122] varies depending on the applicable main option.

In addition to the description of the functionality of each main option given in Chapter 2, it is necessary to look in detail at how the pricing and payment procedures are typically managed on a building project.

Architects acting as PM should note the two distinct actions of assessing the amount due at each assessment date (core clause 50.1) and certifying a payment within one week of each assessment date (core clause 51.1).

---

122 Price for Work Done to Date.

Design

Design responsibility

> The *Contractor* designs the parts of the *works* which the Scope states the Contractor is to design.[123]

This deceptively simple clause is a powerful tool in terms of legitimate placing of design responsibility. Essentially, it is for the architect to decide exactly which elements of the Project should be designed by whom. In practice, this decision-making process requires a clarity of vision in terms of who is best placed to carry that design responsibility, and the Architect should be in a position to exercise their judgement on this in an impartial manner.

The ability to allocate design anywhere between 0% and 100% as between Client and Contractor without any change in the form of contract, or procedures, is advantageous in practice. The only area for caution in this simple but effective approach is in the clarity of explaining both the required design proportion and any cut-off point.

Design submission and acceptance

> The *Contractor* submits the particulars of its design as the Scope requires to the *Project Manager* for acceptance ...

> The *Contractor* does not proceed with the relevant work until the *Project Manager* has accepted his design.[124]

'Future' design

Under many conventional standard form contracts, there has developed a culture of allowing relatively important decisions to be made retrospectively relative to contract formation. This is usually driven by a desire to accelerate into the construction phase when the design phase is still incomplete.

Given that managing projects in real time is a key objective of NEC4, the Contract deliberately omits procedures which allow procrastination. Provisional sums are conspicuous by their absence in NEC4, and this is entirely intentional. In the context of incomplete Consultant design at the point of wishing to appoint the Contractor, the Architect or design team has the choice either to postpone entering into the building contract until that design is complete or to legitimately allocate that design responsibility to the Contractor, or a specialist Subcontractor to the Contractor, and control the quality of that design by means of an appropriate performance specification.[125]

---

123 Core Clause 21.1.
124 Core Clause 21.2.
125 Documented in the Scope.

Temporary works design

Architects will be interested to see that there is a mechanism for dealing with temporary works design.[126]

Design liability

The standard of care usual for designers practising under English law is 'reasonable skill and care', rather than 'fitness for purpose'. Conventionally, design and build contracts either expressly state a reasonable skill and care obligation or imply it by specific comparative reference to other designers.[127] The NEC4 core clauses are silent on design liability, and the default liability therefore must be construed as fitness for purpose. This must be understood in the context of the international aims of NEC4 and the peculiarity of 'reasonable skill and care' to common-law jurisdictions.

In practice, the fitness for purpose obligation may be modified by two methods:

- describing the design liability accurately in the Scope

- invoking Secondary Option X15[128] (Clause X15.5).

Defects

Defects are the domain of the Supervisor, not the Project Manager, which requires attention to be paid to the distinct role – even if the two roles are being undertaken by the same firm of consultants, and even if those consultants are additionally performing a design role.

A Defect is defined as:

- a part of the works which is not in accordance with the Scope

- or a part of the works designed by the *Contractor* which is not in accordance with the applicable law or the *Contractor's* design which the Project Manager has accepted.

Architects will be interested to note the distinction between the *defects date* and the *defects correction period*. The former is the familiar contractual longstop for post-completion defects to be rectified under the contract (without needing to be treated as a breach of contract); the latter is the period within which any individual defect must be corrected, following its notification. In practice, this is a valuable tool for architects to ensure that defects cannot blight the occupation of a new building by being left for long periods before correction. In deciding appropriate periods for individual building projects – to be completed in Contract Data Part One – it is clearly necessary to consider lead-in times which may be necessary to obtain spare parts, etc. in order to correct defects.

---

126 Core Clause 23.1.
127 E.g. JCT Design and Build Contract
128 The *Contractor's* design.

Dispute management

Early warning

The first tier of dispute management under NEC4, beyond the obvious informal level of discussion, is the early warning procedure.[129] The intention is that any matter which may affect time, cost or quality issues is notified to the other party for joint resolution. If necessary, an early warning meeting will be held to agree on a strategy to remove or mitigate the risk. It should be noted that compensation events are not necessarily early warning matters and that notifying an early warning matter is not inextricably linked with notifying a compensation event.

Adjudication

The second tier of dispute management is resolution by adjudication (Option W1 or W2) or a recommendation by the Dispute Avoidance Board if Option W3 has been chosen.

The Tribunal

The final tier of dispute management is by either arbitration or litigation, depending on which Tribunal has been entered into Contract Data Part One.

Communications

Rigour of communication

There are a number of defined forms of communication, in defined directions between specific people, which are intended to be used as clear and unequivocal statements of both the legal and management status of events occurring on the Project. Some users have commented that they experience an increase in paperwork with the NEC; this must, however, be seen in the context of the usual large volume of 'undefined' letters on projects under other forms of contract. In practice, most users find that some system of consecutively numbered proformas[130] for each communication type works well, avoiding the need for contract administration letters completely. The discipline of communications following the NEC procedures requires rigour and facilitates clear, objective statements, rather than subjective letters, which might be open to interpretation.

Collective responsibility

Many of the obligations to communicate are potentially reciprocal between the parties, for example, a notification of an early warning can be either from Project Manager to Contractor or from Contractor to Project Manager.

---

129 Core Clause 15.
130 Proforma examples can be found in various publications relating to the NEC.

Communication types

Architects will be very familiar with instructions and certificates as communications types under building contracts. The additional communication types with which architects will need to become acquainted under NEC4 contracts include notifications, acceptances and records.[131]

Change control

There are some fundamental issues in the context of change control, which may be summarised in the following manner:

- It is pointless to fight change over the lifecycle of a building project; it is inevitable and must therefore be managed.

- Changes over the lifecycle of a project can be few or many and major or minor; their management needs to be appropriate.[132]

NEC4 provides a detailed mechanism for managing change in Core Clause Section 6: Compensation Events. This is probably the most complicated section in the entire NEC4 contract, but also a critical section in terms of enabling efficient project management.

NEC4 adopts a prescriptive approach to change management, which is regarded by some as controversial and by others as revolutionary. The compensation event procedure comprises the following:

1. **Definition**

   There is a list of items which are compensation events.[133] Architects will be used to this approach from other standard form contracts[134] and administering the compensation event procedure will initially be a matter of checking what is happening on a real project against that list.

2. **Notification**

   There is a reciprocal duty for the Project Manager and the Contractor to notify each other of any compensation event when they become aware of it – the detailed obligations are set out in Core Clause 61.

3. **Quotations**

   The Project Manager must instruct the Contractor to prepare a quotation, or alternative quotations, when the compensation event is instigated on behalf of the Client, whether the Project Manager or Supervisor simultaneously gives an instruction, or whether a decision to instruct will depend on the outcome of a quotation. The Project Manager must also instruct the Contractor to prepare a quotation for any other compensation

---

131 See Appendix for Communication checklist by clause number, type and initiator (p. 94–102).
132 Engineering projects will typically have relatively few, but major changes; building projects are likely to have relatively many, but minor changes.
133 Core Clause 60.1.
134 E.g. relevant events under JCT contracts.

event. The timescales both for the Contractor to submit a quotation and for the Project Manager to reply are prescriptive.[135] They can be extended by agreement between the Project Manager and the Contractor,[136] although such agreement should only be based upon genuine necessity.

4. **Assessment**

Compensation events are assessed relative to Defined Cost and the Accepted Programme, and quotations must deal with both time and money. The detailed provisions in Core Clause 63 must be followed by the Contractor. There are circumstances in which the Project Manager must assess a compensation event – these are governed by the detailed provisions in Core Clause 64.

5. **Implementation**

Compensation events are implemented when their time and money implications are set. Architects should note that these implications are set once and only revisited in accordance with the *conditions of contract*.[137]

Time and money are inextricably linked through the programme and Defined Cost.

Architects may consider the compensation event procedure to be particularly challenging; nonetheless, it is worth remembering that conventional standard form contracts have not avoided the need to manage change, they have simply pushed much of the work to the post-completion phase. NEC4 requires contemporaneous change management effort; the reward is to the whole project team in terms of certainty of outcome and avoidance of potential post-completion disputes.

Time-barring

The compensation event procedure includes temporal obligations with respect to change management that are both reciprocal and onerous. The potential time and money rights of the parties in relation to change can be time-barred if the contractual timescales are not followed. Specifically, if the following actions which confer those time and money rights are not taken within the specified time periods, then those rights are extinguished:

- The Contractor must notify a compensation event within eight weeks of becoming aware of it – it is a large carrot to notify a compensation event on time if the right to it can be extinguished after eight weeks.[138]

- The Project Manager must notify the Contractor within two weeks of receipt if any event is considered not to be a compensation event – it is an equally large carrot to check a compensation event on time if it is deemed to be accepted after two weeks by default, i.e. if it is not proactively rejected if it is considered incorrect.[139]

---

135 Core Clause 62.3.
136 Core Clause 62.5.
137 Core Clause 66.3.
138 Core Clause 61.3.
139 Core Clause 61.4.

The purpose of such time-barring is effectively to incentivise the parties to perform efficiently and deal with compensation events as they arise. This is a critically important objective of NEC4 in dealing with change contemporaneously and not leaving it for a final account debate.

Completion

Definition of Completion

The first thing to clarify with completion under NEC4 is that there is no 'practical completion' as found under many other standard form building contracts. The default position is therefore that Completion occurs when the project is actually finished.

The default position can be modified as a result of the content of the Scope and architects should carefully consider the project programme in order to determine whether anything could, or should, be left until after Completion, for example, soft landscaping which ought to take place during a planting season.

Take over

NEC4 provides a mechanism for the Client to use parts of the works prior to Completion, as well as a requirement for the Client to take over the works after Completion.[140] It is open to the parties to regulate the relationship between Completion and the Completion Date at the outset by stating requirements in the Contract Data.

---

140 Core Clause 35.

4 Collaborative working with NEC4

Professional services

Relationship to the building contract

To benefit significantly from the integrated project management principles of NEC4, it will be necessary to bring the NEC4 family together and work as a collaborative team. The first step in this direction is the NEC4 *Orange Book* Professional Service Contract (PSC). There is parity between the PSC *Orange Book* and the NEC4 *Black Book*, such that the contracts operate in a back-to-back manner, with differences only in the conventions applicable to consultants' and contractors' respective roles. The *Orange Book* shares the *Black Book* philosophy, and most of the lessons learned in the context of an NEC4 building contract will be equally applicable to the PSC, including the all-important structure.

This section details the key points which should be considered with the PSC.

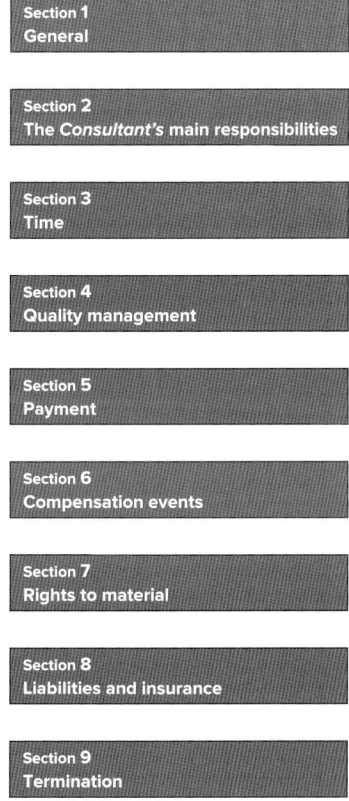

Section 1
General

Section 2
The *Consultant's* main responsibilities

Section 3
Time

Section 4
Quality management

Section 5
Payment

Section 6
Compensation events

Section 7
Rights to material

Section 8
Liabilities and insurance

Section 9
Termination

Figure 21: PSC Core Clauses

Application of the PSC for any discipline of Consultant

All types of professional services can be successfully performed by Consultants appointed under the PSC and on projects requiring a number of different Consultants, having everyone signed up on the same basis can be of enormous benefit.

Cultural and procedural change

There is a need for a cultural shift in perspective to operate the PSC successfully and to reap the rewards. If people cling to old habits and are suspicious of new ideas, then many of the benefits of the PSC will never be realised. However, rapid change and new technologies are now a part of life and learning a new system which is truly new, rather than just new to the user, has the advantage that there has been little time for any bad habits to develop. The objectivity of the PSC also makes it unlikely to be open to many varying interpretations.

The PSC encourages, even demands, better documentation and more efficient management skills. It is often said that there is nowhere left to hide; nevertheless, if the procedures are ignored or incorrectly applied, the error is virtually immediate and can therefore be rectified for the future, rather than potentially handicapping the Project. If a party persistently applies the procedures incorrectly, or even fails to apply them at all, at least a clear breach becomes apparent, instead of arguing over semantics. It clearly takes both parties to 'act in a spirit of mutual trust and co-operation'!

Responsibility, authority and people organisation

The PSC necessitates a clear line of both delegated responsibility and authority. The consequences of failing to provide this are that decisions would be unlikely to be reached within the required timescales. Some consultancy organisations have trained staff on trial projects under the PSC for measurement against projects which have been carried out under conventional discipline-specific appointment documents.

Using NEC4 may be easier if there is a clear understanding of people's functions during pre-contract, post-contract and post-completion stages, which may overlap.

Multidisciplinary and project-specific nature of the Scope

The Scope in the *Orange Book* has the same status as in the *Black Book*. The Contract definition is as follows:[141]

The Scope is information which either
- specifies and describes the *service* or
- states any constraints on how the *Consultant* provides the Service.

---

141 Core Clause 11.2 (16).

The Scope should be arrived at based on building up an appropriate range of services relative to a particular Consultant's discipline and required input on a particular project. While any formulaic tendencies based on ticking 'standard services' should ideally be avoided, there is no difficulty in using discipline-specific conventions as a framework for the Scope.

For example, the RIBA Plan of Work can be used as the framework for defining the range of services required from an architect on a particular project; however, the actual tasks to be performed under each work stage should be carefully considered and documented in the Scope.

If this approach is taken in arriving at the Scope for each Consultant, it should ensure that there are no gaps or overlaps in the total range of Consultants' services performed on a project, providing a back-to-back seamless range of professional services for a project.

Consultant designers in particular have conventionally worked to standard services schedules, which do not always assist clients in understanding the complexity of work being undertaken. One of the benefits of the PSC is that it encourages a distinction between tasks being performed and deliverables resulting from such activities, which can be seen as equally helpful to consultants and their clients.

The documentation of the services to be provided under a PSC takes an appropriate form relative to the discipline and the Project, irrespective of whether the Client of the Consultant has conventional client, contractor, or 'lead' consultant status. The PSC, the Professional Service Short Contract (PSSC) or the Professional Service Subcontract (PSS) can be chosen accordingly.

No conventional percentage fee basis

While conventional standard form professional services contracts have been largely based on a percentage fee calculation, and resource-based fee calculations can still be seen as rare in some sectors, the PSC offers considerably more flexibility and accuracy in fee calculation with its main options. Many building clients will no longer tolerate a fee basis predicated on the principle of fees increasing with construction costs; i.e. almost a disincentive for consultants to proactively keep projects within a set budget.

Architects will be interested in the fee bases offered under the PSC payment mechanisms.

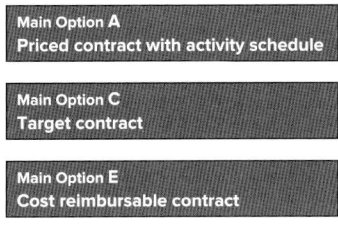

Figure 22: PSC Main Option Clauses

Fine-tuning a PSC Consultant's role

In addition to the complementary main option payment mechanisms, there is also parity in the secondary options as between the PSC *Orange Book* and the NEC4 *Black Book*, with adjustments made to recognise the PSC Consultants' role.

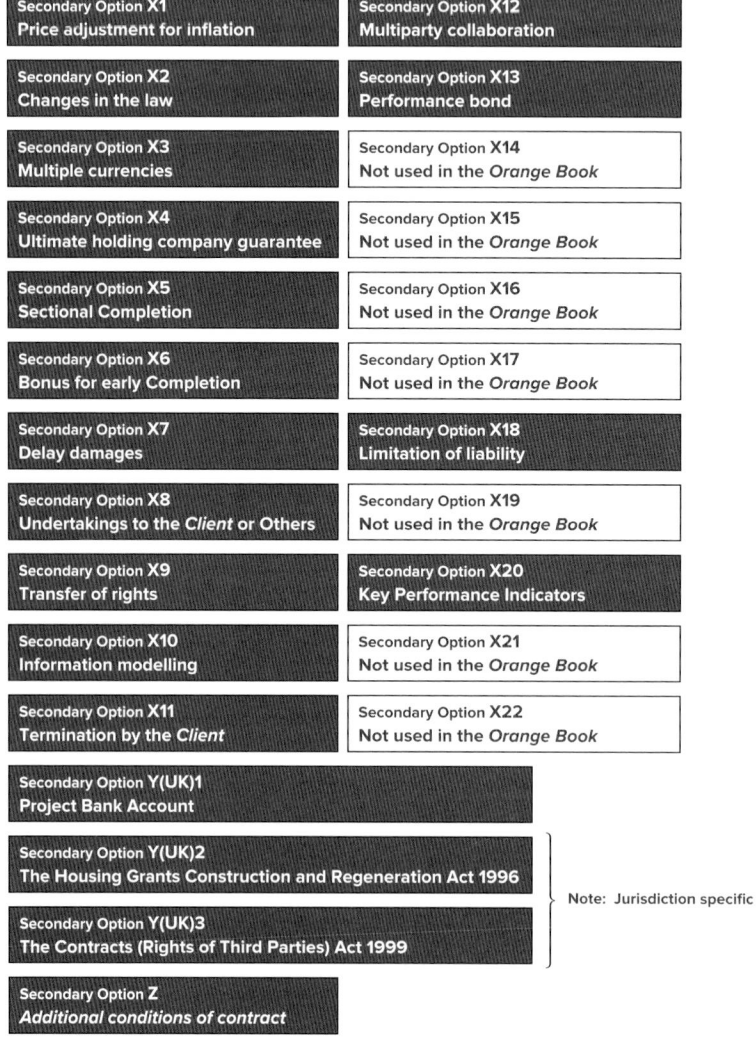

Figure 23: PSC Secondary Option Clauses

Subcontracting

The next step in the direction of integrated project management principles of NEC4 is the NEC4 *Purple Book* subcontract. Again, there is parity between the NEC4 *Purple Book* and the NEC4 *Black Book*, such that the contracts operate in a back-to-back manner, with the difference in the 'step down' so that the Client becomes the Contractor and the Contractor the Subcontractor. Under the NEC4 subcontract conditions, the Contractor manages the time, cost and quality subcontract administration and, in the context of quality matters, there is recognition of both the Client's and the Supervisor's status under the NEC4 *Black Book*.[142] The now familiar NEC4 structure is maintained.

The NEC4 *Purple Book* subcontract main option clauses mirror those of the NEC4 *Black Book*, with the exception of the management contract option (*Black Book* Option F).

A critical point for architects who have attempted to manage subcontract risk to clients under older style standard form contracts or subcontracts is the clarity with which such risk is placed on the Contractor under the NEC4 *Black Book*:

> If the *Contractor* subcontracts work, he is responsible for Providing the Works as if he had not subcontracted. This contract applies as if a Subcontractor's employees and equipment were the *Contractor's*.[143]

Subcontracting under NEC4 will sensibly be carried out under either the *Purple Book* subcontract or, in instances where the subcontracted works are of low complexity, the *Mauve Book* Short Subcontract. Indeed, the NEC4 *Black Book* building contract foresees an NEC contract as the default for all subcontracts unless agreed otherwise.[144]

> The *Contractor* submits the proposed subcontract documents, except any pricing information, for each subcontract to the *Project Manager* for acceptance unless
> - the proposed subcontract is an NEC contract which has not been amended other than in accordance with the *additional conditions of contract* or
> - the *Project Manager* has agreed that no submission is required.

The NEC4 *Purple Book* subcontract secondary option clauses mirror those of the NEC4 *Black Book* entirely.

---

142 NEC4 Subcontract Core Clause 40.
143 NEC4 *Black Book* Core Clause 26.1.
144 NEC4 *Black Book* Core Clause 26.3.

Project profiles

It is extremely important in using NEC4 to its best advantage to properly analyse the profile of individual projects. Such analysis needs to extend beyond the essential procurement strategy and encompass an assessment of participating organisations and the staff employed within them. It would be naive to expect truly collaborative working between organisations to succeed without considering the people who can collectively make it happen – personalities matter!

Architects are often in a pivotal position with regard to fostering collaboration. This position results partly from the leading role architects may take at the outset of projects, including advising clients on the need for other consultants, and partly from an expectation that most architects can offer excellent communication skills.

Truly collaborative working requires a partnering style contract such as NEC4 as a prerequisite for success; however, it would be a fallacy to assume that merely choosing NEC4 as a contractual basis offers any guarantee of success. The people involved in a particular project have to buy into the contractual ethos for a project to be run successfully on a collaborative basis. While highly motivated project teams can achieve some degree of collaboration *in spite of* adversarial contract conditions, it follows that unwilling team members can fall out and jeopardise a project *even with* cooperative contract conditions.

Partnering

Partnering as a concept has been recognised in the building industry for a long time, essentially putting the success of the Project at the forefront of all parties' goals. Partnering has historically had more than one guise. In its simplest form, partnering was an intuitive process between project team members (with cooperative personalities). As a more conscious decision, project teams volunteered to partner by adding a non-binding partnering charter to their (often still adversarial) contract conditions.[145] In its most powerful form, partnering has become mandatory, based on cooperative obligations in a binding contract. It is this latter model to which NEC4 belongs.

---

145 E.g. JCT 98 Practice Note 4 (2001, RIBA Publications).

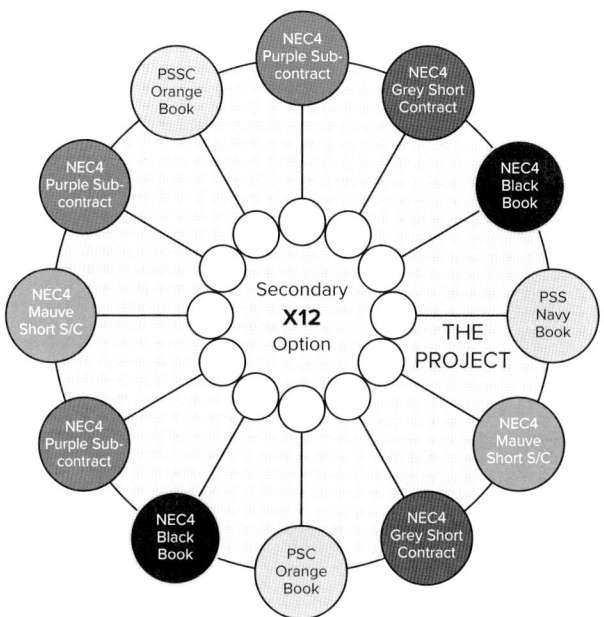

Figure 24: NEC4 Multiparty Collaboration

The history of Secondary Option X12

In the year 2000, the CIC[146] published the findings of their multidisciplinary Task Force research into partnering and provided a possible methodology for contractual partnering.[147] This included model heads of terms for guidance to the construction industry.[148] The NEC panel then carried out a review of NEC compliance with the CIC Guide and decided to retain the family structure of bi-party NEC contracts, including the core clauses, main option and secondary option clauses, as it was considered that the NEC family already incorporated most of the objectives of the CIC Guide. In addition, it was decided that a more formal layer of partnering rights and obligations should be added to NEC at secondary option level. The *White Book* consultative version of the NEC Partnering Option X12 was published as a supplementary NEC document in September 2000, followed by the first edition in June 2001. Secondary Option X12 was incorporated into the main secondary option documentation in NEC3 (2005). Secondary Option X12 was updated and renamed 'Multiparty collaboration' on publication of NEC4 in 2017.

---

146 Construction Industry Council.
147 CIC (2000) *A Guide to Project Team Partnering*, London.
148 Upon which the partnering contract PPC2000 was subsequently based.

Extent of partnering

Secondary Option X12: Multiparty collaboration maintains the flexibility of bi-party contracts but allows specific aspects of partnering to become a contractual obligation if and when required. Contractual partnering arrangements can be tailored to suit both project and 'people' requirements:

- Bi-party partnering – requires an NEC contract.

- Multiparty partnering – requires NEC contracts, including Secondary Option X12.

- Multi-project partnering – requires an NEC contract or series of NEC contracts and can include Secondary Option X12 where multiparty partnering relationships are desired.

Structure and status

Secondary Option X12: Multiparty collaboration sits at secondary option level within the NEC 'pick-and-mix' framework and is incorporated in exactly the same way as other secondary options. It will be incorporated into as many bi-party contracts as required, which in practice will be the NEC4 contracts between all the key participants in a project. In comparing Secondary Option X12: Multiparty collaboration with other standard form contract approaches to partnering, architects should note the following:

- Secondary Option X12 is not a stand-alone contract; it is predicated on an NEC4 family base contract – NEC4 *Black Book*, Subcontract, PSC etc.

- Secondary Option X12 does not create a multiparty contract; all Partners share common bi-party rights and obligations.

- Secondary Option X12 does not create a legal partnership across bi-party contracts.

Secondary Option X12: People definitions and scope of application

There are some additional definitions in Secondary Option X12 to become familiar with:

- Promoter: project sponsor and Client under some bi-party contracts

- Partner: any team member with Secondary Option X12 in their bi-party NEC contract

- Core Group: group of key partners.

Secondary Option X12: Multiparty collaboration can be incorporated at any level in the supply chain; the contribution of a Partner is more important than their size.

The contractual partnering relationship commences with the first bi-party NEC family contract which includes the Secondary Option X12: Multiparty collaboration.

There is provision for Partners to join and leave at appropriate stages of a project or series of projects.

Implementation of Secondary Option X12 documents

The Schedule of Partners is maintained by the Core Group.

The Secondary Option X12 Contract Data identifies the Client and their objective.

The Partnering Information should contain only information which is relevant to all bi-party contracts, i.e. it must be common and not vary across the Partners' contracts. In practice, this means that in compiling the Scope for a building contract and for a professional services contract where Secondary Option X12 is to be incorporated in the contract conditions, architects should ensure that the drafting is done on a side-by-side basis to ensure the required commonality.

Partners' management responsibilities

Secondary Option X12: Multiparty collaboration creates additional rights and obligations to assist in managing the relationship between Partners. The intended live nature of the Partnering Information necessitates continuous review relative to both the Promoter's objective and the other Partners' objectives.

There is an important role for the Core Group in both preventive dispute avoidance and as the first tier of a problem-solving hierarchy. Formal dispute resolution and legal remedies are consciously left within the discrete bi-party contracts. This is based on the principle that the potential embarrassment of having to pass a dispute up the supply chain and back down again, with the certain knowledge of the ultimate Promoter, acts as an active incentive to most bi-party contract parties to resolve their differences amicably. The ultimate sanction for failing to act cooperatively remains, of course, the absence of future partnering projects.

Secondary Option X12: Multiparty collaboration clauses

X12.1 – Identified and defined terms

Partners are named in the Schedule of Partners and include the *Promoter*.

'Own Contract' means the bi-party contract between two Partners incorporating Secondary Option X12.

The Core Group is the listed group of Partners who will steer the partnering relationships on the Project.

Partnering Information specifies how the Partners work together.

Key Performance Indicator targets are stated in the Schedule of Partners.

X12.2 – Actions

The intention is to achieve the *Promoter's* objective as stated in the Secondary Option X12 Contract Data and the other Partners' objectives as stated in the Schedule of Partners.

The Partners select the Core Group and each nominates a representative.

The Core Group's decision-making remit is stated in the Partnering Information. It works democratically and will normally be led by the *Promoter's* representative.

The Core Group maintains the Schedule of Partners and Schedule of Core Group Members throughout the Project.

X12.3 – Collaboration

Collaborative team working is required as stated in the Partnering information, in a 'spirit of mutual trust and co-operation'.

Allowance is made for reciprocal provision of Partners' Information.

There is provision for each Partner to give early warning of matters affecting other Partners' objectives.

The Partners' contributions are coordinated by the Core Group in the form of a timetable, which is incorporated into each Partner's Own Contract programme.

X12.4 – Incentives

Performance of the partnering team, a group of Partners or an individual Partner can be rewarded where a target stated for a Key Performance Indicator is achieved or improved upon.

A final point which architects and their clients will note when comparing partnering methodologies and standard form collaborative contracts is that Secondary Option X12 does not envisage the necessity for an externally appointed partnering advisor. The key reason for this is that individual contracting partners maintain their autonomy with Secondary Option X12, rather than effectively joining an overarching entity as would be the case where a multiparty contract is created.

Framework agreements

Project-specific emphasis

The flexibility of the NEC family allows emphasis to be placed on project-specific solutions. Particularly in the context of long-term relationships and partnering, the NEC family is ideally suited to being used as the basis for an umbrella framework agreement, with subsidiary 'call-off' orders. The precise way in which this is done will depend on the nature of project requirements and the consequential choices of NEC4 family contracts.

Many architects have gained experience of framework agreements in the context of European Union-derived procurement legislation concerning publicly funded – in whole or in part – construction projects. Such framework agreements may have been contractually complex and may have involved bespoke drafting. One of the fallacies which seemed to pervade framework agreements using NEC contracts in the early years of NEC was the apparent presumption that the NEC contracts were invoked at secondary level, underneath a bespoke umbrella contract. This seems to have had no compelling legal or management basis, but rather to have been a hangover from the manner in which older style standard form contracts might be linked together to form a more collaborative framework agreement. A more successful model in the context of NEC would seem to be to capitalise on the back-to-back drafting and introduce an NEC4 contract as the head framework agreement, with further NEC4 contracts nested into it to deal with individual contractual requirements, whether they be on a task-by-task or project-by-project basis.

There are several relatively high-profile project programmes which were developed during the currency of the NEC second edition and were procured on the basis of framework agreements predicated on NEC contracts. A public sector example of these would be one of the NHS procurement programmes for new hospital buildings.[149] A private sector example would be a nationwide expansion of a mobile phone network through additional mast construction.[150]

The NEC Framework Contract

The NEC Framework Contract was introduced into the NEC family with the third edition (2005) and it forms part of the NEC4 publication in 2017. It achieves clarity in providing a specific head contract into which various NEC4 contracts can be nested in order to realise projects.

Term Services

The NEC4 Term Service Contract (TSC) and the Term Service Short Contract (TSSC) are of particular interest in the context of whole life and maintenance contracts. Where appropriate, suppliers can be appointed under the TSC, nested into the NEC4 Framework Contract.

---

149 ProCure 21 – based on an NEC Main Option C contract strategy.
150 Based on an NEC Short Contract strategy.

Management systems

Paperless methods

The corollary to logical and efficient communications under NEC4 contracts is that they lend themselves to being managed under a smart IT system. It is entirely feasible to monitor periods of reply for a variety of communications on a project by means of computerised management systems, which will alert users to periods that are about to expire and enable them to meet imminent contractual deadlines. While such management systems would be complete overkill on smaller or more straightforward projects, there is a plausible argument in their favour on larger or more complex projects. This is particularly the case where a number of NEC4 contracts are interrelated, or nested, with consequentially cascading communication timescales. Generic and NEC-specific software has been developed which assists in managing NEC4 projects, especially in relation to contractual communications, such as notifications.

Many NEC projects have adopted an electronic information-sharing portal for the project team members to exchange information, including contractual communications. The new Secondary Option X10: Information modelling introduced on publication of NEC4 in 2017 sets out a clear basis for managing projects using BIM (Building Information Modelling) and should be helpful to architects.

Support

Architects will probably only want to progress to sophisticated electronic management systems when they have mastered the basics of NEC4 and have honed their communication management skills. There is quite a lot of support available either directly or indirectly from the NEC publishing body, Thomas Telford Ltd. There is also a well-established group to promote the exchange of information about the NEC in use – the NEC Users' Group, which provides a very useful and transparent forum. Finally, there is a website dedicated to NEC and its wider application,[151] which is an obvious starting point for architects, allied professionals and their clients who might be new to NEC4.

---

151 www.neccontract.com

5 International use

Domestic and cross-border

Domestic

Domestic international use means that NEC4 is used by a Client based in a country other than the UK for a project in that country, using consultants/subconsultants and contractors/subcontractors who are also based in that country. In that instance, NEC4 would effectively replace any other standard form contract which might be prevalent in that particular country. While this might seem a perverse thing to do in a country which has an established range of standard form contracts, it would still potentially offer an alternative approach with the ability to marry up legal and project management requirements. In countries that have historically had little or even no choice of standard form contracts, NEC4 offers a blueprint which could form the basis of a national standard form of contract for that emerging construction economy.

Cross-border

Cross-border international use means that NEC4 is used as a neutral contract between parties who have different nationalities and where the Project may be based either in the same country as one of the parties or possibly in a third country. This type of cross-border use is quite likely on larger projects, for example:

- technically demanding projects where word-class expertise can be drawn from an appropriate country

- projects in developing countries where some of the funding may be in the form of international aid.

These concepts of NEC4 international use have already been explored by international parties from countries as diverse as China, South Africa, Germany, Ethiopia, Ireland and New Zealand. The list of interested countries has grown slowly but steadily and it is to be anticipated that the international use of NEC4 will continue to spread around the world.

Jurisdiction

NEC4 is operable in any jurisdiction by virtue of the relatively simple device of removing jurisdiction-specific requirements from the core clauses. The choice of jurisdiction is made by entering the appropriate country into the Contract Data (Part One) under the General section:

The *law of the contract* is the law of XXX

Where XXX would normally be 'England and Wales' for projects in, for example, London.

While it would be unusual, it is also theoretically possibly to decide upon a different jurisdiction from that of the country in which the Project is situated. For instance, if a German bank wished to construct a new office building in London, possibly using some consultants and contractors from Germany, it would be a simple matter of filling in XXX to be 'Germany' in the Contract Data for the jurisdiction of the Project to be German law.

Architects should note that there is no default jurisdiction under the Contract; it follows that it would be a somewhat fatal mistake to fail to fill in the appropriate country in Contract Data Part One in order to designate the law of the Contract.

Architects working in multi-jurisdictional countries such as the UK or the United States should take care in completing this statement in Contract Data Part One. For example, if a Newcastle-based practice were to win a commission in Edinburgh, it would be very important not to just copy the Contract Data entry from a previous project, but rather to have a proper debate with the Client and any other interested parties as to whether the applicable law should be the law of England and Wales, or the law of Scotland?

This capability to choose jurisdiction is somewhat analogous to the situation in international dispute resolution.[152]

Having established the law of the Contract, there is a further key area where architects need to be proactive in ensuring that the national legislative requirements for a project are properly covered. As a direct result of removing jurisdiction-specific requirements from the core clauses, there is no default set of contractual obligations in relation to national legislation, for example, CDM[153] compliance requirements on an English project. This may seem slightly irritating to architects who only ever work on projects under a single jurisdiction, as they could be forgiven for considering it to be an unnecessary additional burden to have to ensure that such legislation is actively referred to.

Quite apart from a real risk of oversight, there is also the risk of not knowing where to add this legislative information to make it as enforceable as if it had been within the generic contract conditions, as it would be with many standard form building contracts. Even for those architects who are not persuaded by this structure for the greater good of ease of international use, there is still a viable alternative to simply choosing an alternative fully nationalised standard form contract. The key is to be found in the all-powerful Scope: this is the correct place to reinstate the requirement to comply with national legislation. Using the example of CDM compliance requirements on an English project, the statement in the Scope could be as general as, 'The Construction (Design and Management) Regulations 2015 must

152 Conflict of Laws.
153 The Construction (Design and Management) Regulations 2015.

be complied with'. Equally, there could be detailed stipulations regarding their specific applicability to the project in question and there could be further contractual requirements outside the strict remit of the legislation, for example, there might be a statement that 'all O&M[154] manuals must be provided both in electronic PDF format and as a hard copy on min. 100 g paper for the purposes of longevity'.

This is clearly an area which requires careful consideration of appropriate national legislation and there is no getting away from the potential gravity of failing to include certain requirements on individual projects. Some architects may immediately wish to revert to the tried and tested approach of older style standard form building contracts in relation to national legislation. However, there is undoubtedly a counterargument to that stance: as the trend within English law appears to be moving towards greater influence of legislation over parties' freedom of contract, it behoves architects to get to grips with such legislation affecting construction and to become familiar enough with it to easily incorporate it appropriately into the Scope for a given project.

Language

Simple English is used throughout NEC4, avoiding the legalistic language usually associated with construction contracts. Subjective statements are avoided and, instead, objective, measurable requirements are stated wherever possible. This approach makes for easier understanding nationally and internationally and has generally been welcomed. Some lawyers commented in the early days of the NEC that the use of the present tense throughout resulted in a lack of distinction between statements of fact and statements of legal obligation; in practice, however, this does not appear to have materialised as a tangible difficulty.

Most first-time users of NEC4 will be struck by the simplicity of the language and will tend to react either positively or negatively – ambivalence is a less likely reaction. Certainly, it would be a mistake to confuse simplicity of language with simplicity of purpose or intent, and experienced users of NEC4 have learned that even the shortest sentences are not without effect.

In the context of international use, there is a strong argument for translating the NEC4 documents into other languages. Nonetheless, in the absence of official translations, the simple English used throughout the documents does assist understanding among those whose mother tongue is not English.

---

154 Operation and Management.

Culture

Cultural diversity is a fascinating aspect of the international use of NEC4, in that its influence can only be established on a trial and error basis. It would never have been possible to assimilate every culture into the drafting of NEC4, as even concentrating on those cultures with a history of using standard form contracts for construction would have required an impossibly large drafting and research body. Instead, NEC4 offers a model that is potentially inclusive of all cultures by means of the following salient devices.

- Necessary disconnection of language from culture

- Necessary disconnection of jurisdiction from culture

- Pick-and-mix contractual structure which can be tailored towards cultural precedents and customs.[155]

The relative success of NEC4 in operating within different and new cultures can only be assessed over decades rather than mere years. Countries outside the UK in which NEC has been used to a significant extent to date include South Africa and New Zealand. While these countries perhaps do not represent the most radically different cultures in the world relative to England, they are useful examples of successful use of NEC in diverse cultures.

The worldwide history of the evolution of standard form contracts suitable for construction projects cannot be divorced from historical and political events, and it is certainly interesting to note how older standard form contracts were initially introduced into other cultures on what could almost have been described as a neo-colonial basis.[156] NEC4 goes a very long way in avoiding the imposition of predetermined and possibly inappropriate cultural provisions, particularly those associated with common-law principles. Nevertheless, given that the NEC contract was born and bred in England, it has to be acknowledged that some English cultural prejudices are bound to remain even after the review leading up to the publication of the fourth edition, NEC4, in 2017. However, to put this into context, even within the confines of the construction industry in England, there are cultural nuances between professions such as architects and engineers, and NEC4 copes well in being flexible enough to be tailored to fit novel briefs.

Overall, NEC4 appears unlikely to present any insurmountable cultural challenges wherever it may be used in the world.

---

155 Including the potential for introducing unique secondary options.
156 E.g. Singapore Standard Form of Building Contract; FIDIC Conditions of Contract (Fédération Internationale des Ingénieurs Conseils).

6 In conclusion: decisive features of NEC4

Relative certainty and *carpe diem*

Temporal longstops and avoidance of delay

The rigour of NEC4 communications coupled with the requirement to fix a period for reply in the Contract Data (Part One) effectively eliminate any significant doubt over when any contractual issue will be resolved. While NEC4 does make provision for consensual extending of timescales,[157] it is reasonable to assume that such consensus will only be forthcoming when it is genuinely in both contractual parties' interest to allow more time for a particular issue. Experience suggests that where NEC4 is properly operated, it is quite rare for communication time periods to be extended. It is perhaps slightly more common on building projects for compensation event quotation and/or reply periods to be extended; this is because the periods of three and two weeks respectively are generically prescriptive,[158] rather than decided on a project-specific basis. Consequently, depending on the building project in question, it is foreseeable that quotations for compensation events may be more time consuming, whether because of the complexity of an individual compensation event, the involvement of subcontractors, or a plethora of contemporaneous compensation events. In such circumstances, it may be in both parties' interests to agree in advance to a limited extension of the period in question, although in order to maintain the relative certainty NEC4 engenders, blanket extensions should never be given.

Real time

One of the most important and potentially beneficial aspects of managing projects under the NEC4 form of contract is the underlying rationale of running projects in real time, without reliance on future negotiations.[159] This can be seen as quite a controversial aspect of NEC4, in that it leaves no room for either procrastination or revisiting difficult decisions. However, this aspect of NEC4 should not be perceived as necessarily requiring greater certainty of project objectives at the outset; indeed, NEC4 is predicated on the principle of needing flexibility.

---

157 Core Clauses 13.5 & 62.5.
158 Core Clause 62.3.
159 *Carpe diem (quam minimum credula postero)* – Seize the day (trusting as little as possible in the future). Horace, born 8 December 65 BC.

One of the best ways for architects to decide whether NEC4 is something they wish to embrace is possibly to reflect on their own approach to management. Anyone for whom decision-making is preferably an incremental process is likely to find the requirements of NEC4 quite onerous. Whereas anyone who tends towards a holistic approach to making the right decision in all the circumstances at a snapshot in time is much more likely to find NEC4 second nature to them.

Consistency

The whole project

The use of NEC contracts to date has seen advantages in appointing everyone within the supply chain for a particular project on an NEC form of contract. These advantages stem from having back-to-back contractual relationships, whether in the context of professional services or in the context of construction. Architects may be used to a convention of seeing the description 'supply chain' as solely relating to contractors and subcontractors; however, given the flexibility of procurement routes which can be supported by NEC4, it is perhaps helpful to get used to the idea that a supply chain includes consultants and subconsultants.

A way of life

Clearly, if NEC4 users get accustomed to and enjoy working collaboratively in such an environment of truly back-to-back contractual relationships, with no gaps or overlaps between the contracts making up the totality of project requirements, it becomes quite appealing to make NEC4 contracts the default for new projects. Architects are in a unique position with their building clients to influence the form of contract proposed for a particular project, in that they are appointed early, if not first, in the process. Of course, there will be other knowledgeable and influential players, such as cost consultants or even contractors, and naturally clients themselves will often have strong views about forms of contract. However, in the context of some architects occasionally feeling sidelined over key project management decisions, such as forms of contract, it is worth remembering that knowledge is power; adequate knowledge of NEC4 will therefore almost certainly put architects in a stronger position to ensure that their designs can be realised reliably and efficiently. It is perhaps also worth remembering that the balance of power in collective knowledge of NEC4 within the construction industry currently rests predominantly outside of the architectural profession and that it is arguably about time that more architects got up to speed. No one is advocating that a particular profession should take sole ownership of NEC4 – it is in essence a democratic form of contract and there are cogent arguments in favour of collaborative ownership, including where appropriate by architects.

Commitment

All or nothing

NEC4 is undoubtedly a standard form of contract which offers the potential to manage projects in a proactive and efficient manner. Indeed, it requires a degree of commitment that is uncommon with other standard form contracts. The Contract is too precise to be operated successfully in anything but a completely committed manner.

Of course, there are real-life examples where the NEC contract has not been particularly well received and where the parties have not benefited from its use; however, there is strong evidence that such an outcome stems primarily from the way in which the parties approach the Contract. It is not a form of contract quite like any of its predecessors and it is certainly not a contract for any ambivalent members of the building industry. If the parties attempt to operate NEC4 in exactly the same way as older standard form contracts or with only half-hearted embracing of the newer approach, they will almost inevitably fail simply because such older standard forms, even in their latest editions, are not predicated on the same level of objectivity or precision as NEC4.

No one is likely to be overheard saying they 'quite like NEC4' – they will invariably tend towards a more polarised position. Those who believe in NEC4 are likely to be people with a 'can do', collaborative approach, who like to have a clear structure for managing a project but who also like a degree of autonomy in making that structure support their particular needs on that project. Anyone who prefers more rigid contractual mechanisms or who has slightly hierarchical tendencies (possibly getting some satisfaction from thrashing out any difficulties on a project with an occasional adversarial 'bunfight') is likely to be less enamoured of NEC4.

In the context of architects advising their clients on procurement strategies, having analysed contract typology and contract form, it may also be beneficial for the players in a building project to indulge in a little psychoanalysis before finalising their project strategy. Depending on the personalities of the key players on a particular project, they may be more or less well suited to the rigours of proactively managing that project under NEC4 contracts.

While there is no absolute necessity to use NEC4 family contracts across the entire supply chain for a project, and there is certainly no prohibition on mixing and matching other standard form contracts with NEC4, it is almost certainly worth actively considering the sense of this on any particular project. The potential success of a matrix of standard form contracts to cover all the relationships on a building project will be as dependent on the personalities of the parties forming those relationships as it will be on the actual standard form contracts. Experience also suggests that even one person out of an entire project team who is actively uncomfortable with a particular contract can be enough to interfere with the operation of that contract to the point where project success is potentially jeopardised.

The moral to this emphasis on personalities and commitment is almost certainly that NEC4 should ideally only be used where the key players have in principle both understood and bought into its philosophy and management techniques.

Use patterns

Use of the NEC contract has multiplied exponentially since its inception. The engineering sector within the construction industry has seen its use become the norm on a range of project types, including highways, railways, water supply and other infrastructure areas. Government endorsement of NEC3 and NEC4 has been important in influencing further use and there is no question that NEC has now been embraced wholeheartedly in the public sector for both engineering and building projects. High-profile projects undertaken using NEC include the London 2012 Olympics and Crossrail, where the NEC integration of engineering and building requirements in the Contract has been replicated in situ with multidisciplinary supply chains. Uptake of the NEC was initially slower in the building sector than the engineering sector, which was probably due to two distinct issues:

- The tendency for engineering works to be on a relatively large scale and sponsored by important, often public sector, patrons, where the incentive is high to adopt state-of-the-art best practice.

- The apparently trivial but disproportionately significant subtitling of the NEC in 1995 (second edition) with 'Engineering and *construction* contract'.

While the latter was clearly intended to emphasise that the NEC philosophy encompasses more than conventional civil, structural or other engineering works, it was arguably too subtle. Had the subtitle been 'Engineering and building contract', it is reasonable to assume that architects and those in their sphere of influence would have picked up on its significance much sooner. Nevertheless, time has taken its course and the building sector has caught up significantly in the use of the NEC form of contract. There are now many buildings standing that have been built under the NEC form of contract, whether hospitals, educational establishments, supermarkets or one-off houses. Public sector clients showed the way and private sector clients have followed.

In the early years of NEC, architects and quantity surveyors or cost consultants tended to be introduced to NEC by their clients, which was perhaps readily understandable. The building sector had a long history of using standard form contracts traceable back to the Victorian era,[160] and architects held a reasonable reluctance to put forward untried or untested methods in any aspect of building. However, a position has been reached where state-of-the-art knowledge in the UK building industry clearly includes the NEC4 form of contract, and architects will therefore want to use it of their own volition and be a 'safe pair of hands' in its implementation.

---

160 JCT Suite of Contracts and its predecessors.

Appendix: NEC4 'Toolkit'

Communication checklist

In accordance with good project management principles, NEC4 communications are intended to cover the entire project and it is therefore no surprise that all the core clause sections include communication requirements.

The range of communications is wide and intended to be exhaustive in relation to the ability to manage a project.

The reciprocal nature of communications under NEC4 can take some practice for architects who are used to administering older style standard form contracts where communication is a little more one way, from contract administering architect to contractor.

The following checklists are intended to give architects for whom NEC4 is a new experience a quick overview of the contractual communications which are foreseen. Most architects will find that mastering these communications is key to successful contract administration under NEC4, and the checklists can still provide a useful aide-memoire to relatively experienced NEC4 users. It's also useful to note that NEC4 includes Example Communication Forms in the Appendices to the User Guides.

Communications BY CLAUSE NUMBER

D = Discretionary Communication / **M** = Mandatory Communication

Action	Clause no.	Communication	Issued By
Notification of agreed extension to *period for reply*	13.5	Notification	*Project Manager*
Instruction to change Scope or a Key Date	14.3	Instruction **D/M**	*Project Manager*
Notification of 'cost/time/quality' early warning matters	15.1	Notification **M**	*PM/Contractor*
Notification of other early warning matter	15.1	Notification **D**	*PM/Contractor*
Instruction to attend early warning meeting	15.2	Instruction **D**	*PM/Contractor*
Revise Early Warning Register to record early warning meeting	15.4	Record	*Project Manager*
Notification of an ambiguity or inconsistency	17.1	Notification	*Project Manager*
Instruction to resolve the ambiguity or inconsistency	17.1	Instruction **M**	*Project Manager*
Notification of illegal or impossible requirements	17.2	Notification	*PM/Contractor*
Instruction as to how to deal with prevention event	19.1	Instruction	*Project Manager*
Instruction to remove an employee	24.2	Instruction	*Project Manager*
Acceptance of Subcontractor	26.2	Acceptance	*Project Manager*
Acceptance of Subcontract	26.3	Acceptance	*Project Manager*
Completion Certificate	30.2	Certificate	*Project Manager*
Acceptance of programme	31.3	Acceptance	*Project Manager*
Notification of non-acceptance of programme	31.3	Notification	*Project Manager*

Action	Clause no.	Communication	Issued By
Instruction to stop or not to start work	34.1	Instruction	*Project Manager*
Take over Certificate	35.4	Certificate	*Project Manager*
Instruction to submit quotation for acceleration	36.1	Instruction	*Project Manager*
Instruction to search for defects	43.1	Instruction	*Supervisor*
Notification of defects	43.2	Notification	*Supervisor/ Contractor*
Defects Certificate	44.3	Certificate	*Supervisor*
Notification of extension to defects correction period	44.4	Notification	*Project Manager*
Acceptance of quotation to accommodate defect	45.2	Acceptance	*Project Manager*
Instruction to change Scope, Prices and Completion Date	45.2	Instruction	*Project Manager*
Assessment of amount of money due	50.1	Assessment	*Project Manager*
Payment Certificate	51.1	Certificate	*Project Manager*
Notification of compensation event	61.1	Notification	*Project Manager*
Instruction to submit quotation for compensation event	61.2	Instruction	*Project Manager*
Notification of failure to give early warning	61.5	Notification	*Project Manager*
Notification of correcting compensation event effect	61.6	Notification	*Project Manager*
Instruction to submit alternative quotations	62.1	Instruction	*Project Manager*
Acceptance of a quotation	62.3	Acceptance	*Project Manager*
Instruction to submit a revised quotation	62.3	Instruction	*Project Manager*
Notification of PM's own assessment	62.3	Notification	*Project Manager*

Action	Clause no.	Communication	Issued By
Notification of extension of quotation period	62.5	Notification	*Project Manager*
Notification of PM's assessment of compensation event	64.3	Notification	*Project Manager*
Instruction to deal with objects of value/historic interest	73.1	Instruction	*Project Manager*
Acceptance of insurance details	84.1	Acceptance	*Project Manager*
Termination Certificate	90.1	Certificate	*Project Manager*
Notification of default	91.2	Notification	*Project Manager*

Communications BY TYPE

D = Discretionary Communication / **M** = Mandatory Communication

Action	Communication	Issued By	Clause no.
Acceptance of Subcontractor	Acceptance	*Project Manager*	26.2
Acceptance of Subcontract	Acceptance	*Project Manager*	26.3
Acceptance of programme	Acceptance	*Project Manager*	31.3
Acceptance of quotation to accommodate defect	Acceptance	*Project Manager*	45.2
Acceptance of a quotation	Acceptance	*Project Manager*	62.3
Acceptance of insurance details	Acceptance	*Project Manager*	84.1
Assessment of amount of money due	Assessment	*Project Manager*	50.1
Completion Certificate	Certificate	*Project Manager*	30.2
Take over Certificate	Certificate	*Project Manager*	35.4
Defects Certificate	Certificate	*Supervisor*	44.3
Payment Certificate	Certificate	*Project Manager*	51.1
Termination Certificate	Certificate	*Project Manager*	90.1
Instruction as to how to deal with prevention event	Instruction	*Project Manager*	19.1
Instruction to remove an employee	Instruction	*Project Manager*	24.2
Instruction to stop or not to start work	Instruction	*Project Manager*	34.1
Instruction to submit quotation for acceleration	Instruction	*Project Manager*	36.1
Instruction to search for defects	Instruction	*Supervisor*	43.1
Instruction to change Scope, Prices and Completion Date	Instruction	*Project Manager*	45.2
Instruction to submit quotation for compensation event	Instruction	*Project Manager*	61.2

Action	Communication	Issued By	Clause no.
Instruction to submit alternative quotations	Instruction	*Project Manager*	62.1
Instruction to submit a revised quotation	Instruction	*Project Manager*	62.3
Instruction to deal with objects of value/historic interest	Instruction	*Project Manager*	73.1
Instruction to attend early warning meeting	Instruction **D**	*PM/Contractor*	15.2
Instruction to change Scope or a Key Date	Instruction **D/M**	*Project Manager*	14.3
Instruction to resolve the ambiguity or inconsistency	Instruction **M**	*Project Manager*	17.1
Notification of agreed extension to *period for reply*	Notification	*Project Manager*	13.5
Notification of an ambiguity or inconsistency	Notification	*Project Manager*	17.1
Notification of illegal or impossible requirements	Notification	*PM/Contractor*	17.2
Notification of non-acceptance of programme	Notification	*Project Manager*	31.3
Notification of defects	Notification	*Supervisor/ Contractor*	43.2
Notification of extension to defects correction period	Notification	*Project Manager*	44.4
Notification of compensation event	Notification	*Project Manager*	61.1
Notification of failure to give early warning	Notification	*Project Manager*	61.5
Notification of correcting compensation event effect	Notification	*Project Manager*	61.6
Notification of PM's own assessment	Notification	*Project Manager*	62.3

Action	Communication	Issued By	Clause no.
Notification of extension of quotation period	Notification	*Project Manager*	62.5
Notification of PM's assessment of compensation event	Notification	*Project Manager*	64.3
Notification of default	Notification	*Project Manager*	91.2
Notification of 'cost/time/quality' early warning matters	Notification **M**	*PM/Contractor*	15.1
Notification of other early warning matter	Notification **D**	*PM/Contractor*	15.1
Revise Early Warning Register to record early warning meeting	Record	*Project Manager*	15.4

Communications BY ISSUER

D = Discretionary Communication / **M** = Mandatory Communication

Action	Issued By	Clause no.	Communication
Notification of illegal or impossible requirements	*PM/Contractor*	17.2	Notification
Notification of 'cost/time/quality' early warning matters	*PM/Contractor*	15.1	Notification **M**
Notification of other early warning matter	*PM/Contractor*	15.1	Notification **D**
Instruction to attend early warning meeting	*PM/Contractor*	15.2	Instruction **D**
Notification of agreed extension to *period for reply*	*Project Manager*	13.5	Notification
Instruction to change Scope or a Key Date	*Project Manager*	14.3	Instruction **D/M**
Revise Early Warning Register to record early warning meeting	*Project Manager*	15.4	Record
Notification of an ambiguity or inconsistency	*Project Manager*	17.1	Notification
Instruction to resolve the ambiguity or inconsistency	*Project Manager*	17.1	Instruction **M**
Instruction as to how to deal with prevention event	*Project Manager*	19.1	Instruction
Instruction to remove an employee	*Project Manager*	24.2	Instruction
Acceptance of Subcontractor	*Project Manager*	26.2	Acceptance
Acceptance of Subcontract	*Project Manager*	26.3	Acceptance
Completion Certificate	*Project Manager*	30.2	Certificate
Acceptance of programme	*Project Manager*	31.3	Acceptance
Notification of non-acceptance of programme	*Project Manager*	31.3	Notification

Action	Issued By	Clause no.	Communication
Instruction to stop or not to start work	*Project Manager*	34.1	Instruction
Take over Certificate	*Project Manager*	35.4	Certificate
Instruction to submit quotation for acceleration	*Project Manager*	36.1	Instruction
Notification of extension to defects correction period	*Project Manager*	44.4	Notification
Acceptance of quotation to accommodate defect	*Project Manager*	45.2	Acceptance
Instruction to change Scope, Prices and Completion Date	*Project Manager*	45.2	Instruction
Assessment of amount of money due	*Project Manager*	50.1	Assessment
Payment Certificate	*Project Manager*	51.1	Certificate
Notification of compensation event	*Project Manager*	61.1	Notification
Instruction to submit quotation for compensation event	*Project Manager*	61.2	Instruction
Notification of failure to give early warning	*Project Manager*	61.5	Notification
Notification of correcting compensation event effect	*Project Manager*	61.6	Notification
Instruction to submit alternative quotations	*Project Manager*	62.1	Instruction
Acceptance of a quotation	*Project Manager*	62.3	Acceptance
Instruction to submit a revised quotation	*Project Manager*	62.3	Instruction
Notification of PM's own assessment	*Project Manager*	62.3	Notification
Notification of extension of quotation period	*Project Manager*	62.5	Notification
Notification of PM's assessment of compensation event	*Project Manager*	64.3	Notification

Action	Issued By	Clause no.	Communication
Instruction to deal with objects of value/historic interest	*Project Manager*	73.1	Instruction
Acceptance of insurance details	*Project Manager*	84.1	Acceptance
Termination Certificate	*Project Manager*	90.1	Certificate
Notification of default	*Project Manager*	91.2	Notification
Instruction to search for defects	*Supervisor*	43.1	Instruction
Notification of defects	*Supervisor/ Contractor*	43.2	Notification
Defects Certificate	*Supervisor*	44.3	Certificate

Index